TRANSFORMS FOR ENGINEERS

Transforms for Engineers

A Guide to Signal Processing

K. G. BEAUCHAMP

formerly Director of Computer Services
University of Lancaster

CLARENDON PRESS · OXFORD

1987

Oxford University Press, Walton Street, Oxford OX2 6DP
Oxford New York Toronto
Delhi Bombay Calcutta Madras Karachi
Petaling Jaya Singapore Hong Kong Tokyo
Nairobi Dar es Salaam Cape Town
Melbourne Auckland
and associated companies in
Beirut Berlin Ibadan Nicosia

Oxford is a trade mark of Oxford University Press

Published in the United States
by Oxford University Press, New York

British Library Cataloguing in Publication Data
Beauchamp, K. G.
Transfoms for engineers: a guide to signal
processing.
1. Signal processing
I. Title
621.38'043   TK5102.5
ISBN 0-19-856174-1

Library of Congress Cataloging in Publication data
Beauchamp, K. G.
Transforms for engineers.
Includes bibliographies and index.
1. Signal processing—Mathematics. 2. Transformations
(Mathematics) I. Title.
TK5102.5.B349 1986   621.38'043   86-16421
ISBN 0-19-856174-1

Filmset and printed in Northern Ireland
at The Universities Press (Belfast) Ltd.

PREFACE

In recent years a problem has arisen for the engineer and others interested in the practical application of domain transformation. This is to decide how to comprehend and assimilate the wide range of new signal-processing techniques presently available for the solution of application problems. It is becoming increasingly difficult to select those aspects of the new, as well as the old, methods to match the project under consideration without having to devote a disproportionate amount of time to an analysis of many alternative lines of approach.

This applies particularly to the choice of transformation method applied to a time or other discrete series. What are the methods currently available? How do they compare? Which is better for the purpose? Where are the essential key papers describing methods and algorithms? Should a hardware or a software solution be attempted? For those not especially mathematically inclined, what are the essential mathematical tools required to understand and apply the new advances?

While not assuming to encompass all the detailed answers required it is hoped that this volume will at least point the reader in the right direction. It aims to provide a guide to signal-processing methods used in a wide variety of applications and to introduce readers to the range and use of the many alternative fast orthogonal-transform algorithms now becoming available. Few detailed mathematical derivations are included and in the limited space available only an outline survey of the principal methods can be given. However it is hoped that in all cases the analysis requirements will be clarified and useful comparisons drawn between different techniques which are relevant to the users' problems. An extensive bibliography is given at the end of each chapter indicating the key and supporting papers for the subject areas considered.

Chapters 1 and 2 provide the basic essentials of digital signal-processing methods including a brief survey of the processes of signal representation, data acquisition, filtering, correlation, and spectral analysis. A range of orthogonal transformations are discussed and an introduction given to two-dimensional transformation. Before fast transformation is considered an introductory mathematics tutorial is given in Chapter 3 which aims to equip the reader with the essential tools of linear mathematics and matrix algebra needed to follow recent developments in discrete transform theory and application.

Characteristics of a range of sinusoidal (Fourier) and non-sinusoidal (Walsh and other) fast transformations are given in Chapters 4 and 5 which consider methods and suitability for a variety of signal-processing tasks. Although actual computer programming of the algorithms falls outside the scope of the book it is hoped that sufficient information will be found to give the reader an insight into the 'building blocks' of such programs so that the merits of existing library programs can be assessed.

Success in many signal-processing applications lies in the correct choice of hardware technology and methods. Chapter 6 is intended to indicate the hardware choices available for transformation and to show where they may be usefully employed. Particular interest is centered on VLSI methods and the programmable signal processor.

The remaining Chapters 7 to 10 are application reviews of a number of signal-processing areas where transformation methods are dominant. These include speech processing, communications, sonar, seismology, radar, biomedicine, and image processing. The emphasis here is on description and selection of methods and includes brief mention of some of the alternative statistical and regressive methods with which the transform techniques may be compared.

The author is most grateful for the assistance given to him by a number of organizations and individuals during the preparation of this book. Special thanks are given to Professor H. F. Harmuth of the Catholic University of America, Dr C. K. Yuen of the National University of Singapore, Dr N. Brooke of Lancaster University, and Dr P. Robinson of Sonar Systems Ltd for many helpful discussions; to Dr Theilheimer of the US Department of the Navy, Professor R. Kitai of McMaster University, Canada, Dr J. O. Thomas of Imperial College of Science and Technology, and Dr Linkens of Sheffield University for permission to quote some of their work; to George Allen & Unwin and to Academic Press for permission to include some of the authors earlier published work and to Texas Instruments Ltd, Advanced Micro Devices Ltd, and the Moor Hospital, Lancaster for illustrative diagrams. Finally the assistance of library staff at Lancaster University and at the Institution of Electrical Engineers, London is gratefully acknowledged.

K. G. B.

Lancaster
October 1985

CONTENTS

ABBREVIATIONS AND SYMBOLS   xiii

1   DIGITAL SIGNAL PROCESSING   1

   1.1   Signal processing   1
        1.1.1   Continuous and discrete signals   2
        1.1.2   Representation of signals   2

   1.2   Data acquisition   5

   1.3   Digital filtering   8
        1.3.1   IIR and FIR filters   9
        1.3.2   Spectral windows   10
        1.3.3   Wiener filtering   13

   1.4   Correlation and convolution   14
        1.4.1   Circular correlation   15
        1.4.2   Correlation with real signals   17

   1.5   Spectral analysis   19
        1.5.1   The periodogram   20
        1.5.2   The Blackman–Tukey method   21
        1.5.3   The segmentation method   22
        1.5.4   Data modelling   22
        1.5.5   Cepstrum techniques   24

   1.6   Two-dimensional processing   24

   References   25

2   ORTHOGONALITY AND DOMAIN TRANSFORMATION   27

   2.1   Orthogonality   27
        2.1.1   Complex representation of Fourier series   29

   2.2   Fourier transformation   31
        2.2.1   Some Fourier transforms   32

   2.3   The discrete Fourier transform   33
        2.3.1   Some properties of the DFT   34

   2.4   Non-sinusoidal functions   35
        2.4.1   Walsh and Haar series   35
        2.4.2   Transformation of Walsh and Haar series   39
        2.4.3   Problems and special characteristics   40

   2.5   Two-dimensional transformation   41
        2.5.1   The cosine transform   44

2.5.2   Walsh and Haar transforms   45

References   46

3 MATRICES AND OTHER ESSENTIAL
   MATHEMATICS   48

   3.1 Introduction   48

   3.2 Linear mathematics   48
        3.2.1   Scalars, vectors, and functions   48
        3.2.2   Orthogonality   51
        3.2.3   Linear operators   51

   3.3 Matrix algebra   52
        3.3.1   Determinants   55
        3.3.2   Unitary matrices   55
        3.3.3   Eigenvectors and eigenvalues   56
        3.3.4   Kronecker products   57
        3.3.5   Matrix factorization   58

   3.4 Some specialized matrices   61
        3.4.1   The Hadamard matrix   62
        3.4.2   The Dyadic matrix   62
        3.4.3   Circulant matrices   63
        3.4.4   Circular convolution   64

   3.5 Discrete data manipulation   64
        3.5.1   Bit reversal   64
        3.5.2   Modulo-2 arithmetic   65
        3.5.3   Gray code conversion   66

   References   66

4 FAST FOURIER TRANSFORMATIONS   68

   4.1 Introduction   68

   4.2 Fast Fourier-transform algorithms   70
        4.2.1   Radix-2 transforms   74
        4.2.2   Matrices and the signal flow diagram   79
        4.2.3   The signal flow diagram   83
        4.2.4   Twiddle-factor algorithms   88
        4.2.5   Inverse transformation   88
        4.2.6   Transformation of real data   89
        4.2.7   Transforms for radix-4, radix-8 and mixed radix 91
        4.2.8   Bit-reversal   92

4.3   Reduced multiplication algorithms   93
      4.3.1   Prime factor algorithms   94
      4.3.2   Short DFT algorithms   98
      4.3.3   The Winograd–Fourier transform   101
References   104

5   OTHER FAST TRANSFORMATIONS   107
    5.1   Introduction   107
    5.2   The fast Walsh transform   107
          5.2.1   Derivation of Walsh-function series   107
          5.2.2   Fast transformation   109
          5.2.3   Phase-invariant transforms   114
    5.3   The fast Haar transform   114
    5.4   The fast slant transform   118
    5.5   The generalized transform   120
          5.5.1   The perfect shuffle   123
          5.5.2   Fast unitary transforms   124
    5.6   Two-dimensional transformation   125
          5.6.1   The fast cosine transform   127
    References   130

6   IMPLEMENTATION   132
    6.1   Introduction   132
    6.2   Hardware transformation   132
          6.2.1   Bit-slice operation   133
          6.2.2   Parallel processing   135
          6.2.3   Pipeline processing   139
    6.3   Microprocessor implementation   141
    6.4   Microtechnology   144
          6.4.1   VLSI transformation   144
          6.4.2   Charge-coupled devices   146
          6.4.3   Surface-acoustic-wave devices   148
    6.5   Programmable signal processors   152
          6.5.1   Modular architecture   152
          6.5.2   Microprogramming the PSP   154
          6.5.3   The 'single-chip' PSP   155
    References   158

7 SPEECH PROCESSING AND COMMUNICATIONS 163
  7.1 Introduction 163
  7.2 Speech processing 164
    7.2.1 A model for speech production 164
    7.2.2 Speech coding 165
    7.2.3 Waveform coders 166
    7.2.4 Adaptive transform coding 167
    7.2.5 Vocoders 168
    7.2.6 Pitch-recognition vocoders 169
    7.2.7 Linear prediction 170
    7.2.8 The homomorphic vocoder 172
    7.2.9 Speech synthesis 175
  7.3 Speech recognition 176
    7.3.1 Digit recognition 178
    7.3.2 Continuous speech 179
    7.3.3 Word spotting 180
  7.4 Data communication 180
    7.4.1 Data compression 183
    7.4.2 Sequency multiplexing 185
    7.4.3 TDM–FDM conversion 186
  References 188

8 SONAR, SEISMOLOGY, AND RADAR 193
  8.1 Introduction 193
  8.2 Sonar 193
    8.2.1 Passive sonar 194
    8.2.2 Sonar arrays 197
    8.2.3 Active sonar 197
    8.2.4 Sonar imaging 200
  8.3 Seismology 202
    8.3.1 Seismic signal analysis 203
    8.3.2 Seismic arrays 205
    8.3.3 Seismic data compression 207
  8.4 Radar 208
    8.4.1 Moving target indication 209
    8.4.2 Transform echo-detection methods 210
    8.4.3 Other methods 212
    8.4.4 Non-sinusoidal radar 212
  References 214

9   BIOMEDICINE   218

    9.1   Introduction   218

    9.2   1-D processing   218
          9.2.1   ECG analysis   219
          9.2.2   EEG analysis   222

    9.3   2-D processing   224
          9.3.1   Radioisotope scanning   225

    9.4   Computer-assisted Tomography   226
          9.4.1   X-rays   226
          9.4.2   Nuclear magnetic resonance   229
          9.4.3   Ultrasound   231
          9.4.4   Emission-computed tomography   232

          References   233

10 IMAGE PROCESSING   237

    10.1   Introduction   237
           10.1.1   Some basic definitions   238

    10.2   Image compression and transmission   241
           10.2.1   Image transmission   242
           10.2.2   Television data compression   244

    10.3   Image restoration and enhancement   245
           10.3.1   Image restoration   245
           10.3.2   Intensity mapping   248
           10.3.3   Edge detection   248
           10.3.4   Pseudo-colour enhancement   251

    10.4   Pattern recognition   251
           10.4.1   Template matching   251
           10.4.2   Decision theoretic approach   253
           10.4.3   Syntactic approach   254

           References   256

SELECTED ADDITIONAL REFERENCES   260

INDEX   263

ABBREVIATIONS AND SYMBOLS

ADM	adaptive delta modulation	IIR	infinite-impulse response
ARMA	auto-regressive moving average	KLT	Karhunen–Loève transform
APCM	adaptive pulse-code modulation	LPC	linear-predictive coding
ATC	adaptive transform coder	LPM	linear-predictive method
CAL	directly-symmetrical Walsh function series	MEM	maximum-entropy method
		MLM	maximum-likelihood method
CT	computer-assisted tomography	MPX	multiplexor
C–T	Cooley–Tukey (algorithm)	MSE	mean-square error
cos	cosine	MTI	moving-target indicator
DCT	discrete cosine transform	NTSC	National Television Systems Committee (USA)
DCR	data compression ratio		
det	determinant	NMR	nuclear magnetic resonance
DFT	discrete Fourier transform	O^2DFT	odd-time odd-frequency discrete Fourier transform
DHT	discrete Haar transform		
diag A	diagonal matrix	PCM	pulse-coded modulation
DPCM	differential pulse-code modulation	PET	positron-emission tomography
		PFA	prime-factor algorithm
DSB	double side-band	PR	pseudo-random
DST	discrete sine transform	PSD	power-spectral density
DSDM	digital sequency-division multiplexor	PSF	point-spread function
		QRS	cycle of cardiac activity
DWT	discrete Walsh transform	RT	rapid transform
ECG	electrocardiograph	SAL	inversely symmetrical Walsh function series
ECT	emission-computed tomography		
EEG	electroencephalograph	SBC	sub-band coder
exp	exponential	sin	sine
FCT	fast cosine transform	SNR	signal-to-noise-ratio
FDCT	fast discrete cosine transform	SSB	single side-band
FDM	frequency-division multiplexing	S–T	Sande–Tukey (algorithm)
FFT	fast Fourier transform	SVD	singular-value decomposition
FHT	fast Haar transform	TDM	time-division multiplexing
FIR	finite-impulse response	USI	ultrasonic imaging
FM	frequency modulation	WAL	Walsh function series
FWT	fast Walsh transform	WFTA	Winograd–Fourier transform algorithm
HAD	Hadamard function series		
HAR	Haar function series	WHT	Walsh–Hadamard transform
IDFT	inverse discrete Fourier transform	1-D	one-dimensional
		2-D	two dimensional
IDWT	inverse discrete Walsh transform		

Symbols

a_k, b_k	Fourier coefficients	A^{-1}	inverse of matrix A
A_k, B_k	complex Fourier coefficients	b	binary digit
A^{T}	transpose of matrix A	B	bandwidth (Hz)

c	velocity of light (3×10^8 m s^{-1})	P	permutation matrix
CT	cosine transform matrix	$P(f)$	power-spectral density
C_{xy}	convolution function	P_k	discrete power-spectral density
C	circulant matrix	$\hat{P}_k$	smoothed power-spectral density
D	diagonal matrix		
D	delay	Re(.)	real value of (.)
e	2.718 (base of Naperian logarithms)	R_{xx}	auto-correlation function
e	error	R_{xy}	cross-correlation function
f	frequency (Hz)	s	second
$f(t)$	continuous function of frequency	s_i	signal coefficient
		$s(x)$	sampling function
f_N	Nyquist frequency	S	slant matrix
f_0	carrier frequency	t	time, time delay
f_s	sampling frequency	T	sampling interval, time
$f(x, y)$	two-dimensional image function	T	transpose
$f_a(x, y)$	two-dimensional sampled image function	u, v	spatial coordinates
		v, V	velocity
F	DFT matrix	w	spectral window
G	filter matrix, generalized matrix	w_i	discrete wavelets
		W	$\exp(j2\pi/N)$
G	gain factor	W	Walsh matrix
h	sampling interval	x_i	discrete sampled data or time series
$h(u, v, x, y)$	PSV for a 2-D system	$x(s)$	sampled analog signal
H	Hadamard matrix	$x(t)$	analog signal
Ha	Haar matrix	$x_{i,k}$	two-dimensional discrete sampled series
Hz	Hertz, cycles s^{-1}	$x_{i,k...M}$	multi-dimensional discrete sampled series
$H(\omega)$	frequency-transfer function	x, y	spatial coordinates
I	unit matrix	X^*	complex conjugate of X
$I_m(\cdot)$	imaginary value of ($\cdot$)	X_C	CAL transform coefficients
j	$\sqrt{(-1)}$	X_f	Fourier transform coefficients
K	Kronecker product matrix	X_n	transformed series of x_i discrete values
kb	kilobit		
K_c	complex cepstrum	$X_{n,m}$	two-dimensional transformed series for $x_{i,k}$ discrete values
K_p	power cepstrum		
n_i	discrete noise coefficient	X_S	SAL transform coefficients
$n(x, y)$	two-dimensional noise function	y_i	discrete sampled data or time series
N	number of values in a series		
$o(x, y)$	two-dimensional object function	Zps	sequency (zero crossings/per second)
p	$\log_2 N$		

Greek symbols

β	angle	θ	angle
δ	Dirac delta function	η	relative bandwidth
$\delta(x - n)$	delta function series	λ	eigenvalues, wavelength
ε	dielectric constant	μ	magnetic permeability

π	3.14159	ϕ	angle, phase
σ	conductivity	ω	$2\pi f$, angular frequency
τ	time delay		

Mathematical symbols

Ⓒ	comparison	↔	transform operator
★	complex conjugate	$|\cdot|$	modulus
⊕	modulo-2 addition	ˆ	estimated value
⊗	Kronecker multiplication		

1

DIGITAL SIGNAL PROCESSING

1.1 Signal processing

In this chapter an overview of the broad topic of signal processing is given. This is intended first to establish the terminology and symbols used in later chapters and second to identify those areas where transform operations can most profitably be carried out.

A signal may be defined as any representation of a physical variable in one or more dimensions which can be stated in electrical form as a collection of waveforms (or a waveform). It is nearly always necessary to restate these waveforms as a set of digital numbers so that computing operations can be carried out upon them. Hence an early processing operation will be sampling and quantization of the continuous process. We may need then to select or compare features found in the waveforms in relation to their position in the time, frequency, or space domains. We need the ability to discard certain sections of the acquired signal in any of these domains without affecting other sections, in order to assist recognition of given physical characteristics. This can imply reduction in data quantity without impairing signal quality. It can also result in a redefinition of a signal to mean those waveforms contained in a data sequence conveying relevant information, while categorizing the remainder as a 'noise' signal to be discarded. Finally we may wish to prepare for visual inspection some of the processed data, or retain these in electronic storage for future access.

Thus signal processing covers a broad spectrum of topics: digitization, digital filtering, domain transformation, spectral analysis, correlation, data compression, signal-to-noise enhancement, data display, and in fact practically all those operations on a signal other than decision-making and interpretation.

The basic principles of digital processing are well documented in textbooks at all levels [1–5]. The purpose of this short introductory chapter is simply to present some essential background information, together with the signal-processing terminology used in order to relate to a more detailed look later at spectral domain transformation which plays such an essential role in signal processing operations.

1.1.1 Continuous and discrete signals

A single-dimensional signal is shown in Fig. 1.1(a), in which $x(t)$ represents a continuous or analog signal as a function of time t. If this is subject to a regular sampling process taking a narrow slice of the signal N times per second then we have a data sequence $x(s)$ where $s = 0, 1, 2, \ldots$. An ideal sampling process provides an output which is a train of impulses as shown in Fig. 1.1(b) and where the impulses have an area, $x(nh)$ equal to the magnitude of $x(t)$ at the sampling instants $t = 0$, h, $2h$, ..., $(N-1)h$, where h is the *sampling interval* and $n = 0, 1, 2, \ldots, N-1$. Here N is the number of data samples involved in the process. The process may be expressed as

$$x(s) = x(t) \sum_{n=0}^{n=N-1} \delta(t - nh). \qquad (1.1)$$

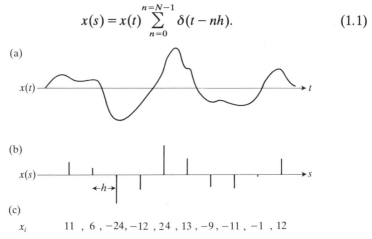

Fig. 1.1 A single-dimensional signal.

The summated term represents a periodic impulse sequence and $\delta(\tau)$ indicates a *Dirac impulse function*, i.e. the limiting case of a pulse of width ε, centred at $t = \tau$, and height $1/\varepsilon$, as $\varepsilon \rightarrow 0$ [6]. Each sample in the signal sequence can have a number associated with it which describes the slice amplitude in digital coded form. In a practical signal we deal with a finite sequence of length N, which is considered zero outside some finite interval. Typically this interval is the first N non-negative integers $0 < n < N - 1$ and the data series generated by this process of sampling quantization and digital coding is expressed as x_i where $i = 0, 1, 2, \ldots, N - 1$.

1.1.2 Representation of signals

We can recognize two broad types of signals; deterministic signals which can be derived explicitly in terms of mathematical relationships and hence contain some sort of cyclic repetition, and random signals, the future values of which cannot be predicted precisely. Nearly all the signals we are likely to meet in physical measurement are random or are associated with signals that are. Analysis of such signals must then proceed in terms of statistical values from which it is possible that a deterministic relationship may be derived. One criterion used to decide whether the signal is deterministic or random is to compare several sets of data obtained under identical conditions over a reasonable period of time. If similar results are obtained then the data are likely to be deterministic.

Periodic deterministic signals can be analysed quite adequately through Fourier series where $x(t)$ is considered to be comprised of harmonically related sinusoids and where each harmonic component repeats itself exactly for all values of t, i.e.

$$x(t) = A_0 + A_1 \sin(\omega_0 t + \theta_1) + A_2 \sin(2\omega_0 t + \theta_2) + \ldots + A_n \sin(n\omega_0 t + \theta_n)$$

(1.2)

or

$$x(t) = A_0 + \sum_{n=1}^{\infty} A_n \sin(n\omega_0 + \theta_n)$$ (1.3)

where A_0 = mean level of signal, A_n = peak amplitude of the nth harmonic, ω_0 = fundamental angular frequency = $2\pi f_0$, and θ_n = phase of the nth harmonic with respect to the fundamental.

Non-periodic or transient deterministic signals are those which decay to a zero value over a finite time interval. For these, an infinite number of frequency components are present in the signal and the appropriate representation is the *Fourier integral transform*

$$X(f) = \int_{-\infty}^{\infty} x(t) \exp(-j2\pi ft) \, dt.$$ (1.4)

This is a complex quantity and can only be represented completely by an amplitude and phase value over the frequency range, although a modulus form $|X(f)|$ is often used for spectral representation (Fig. 1.2).

Random signals require statistical methods for their analysis so that while transformation methods can be applied they will lead to *estimates* or averages of the quantities we seek and the accuracy of these estimates is an important issue in all forms of signal processing with random signals.

A further complication occurs where the signal varies its frequency content with time. We call these signals *non-stationary signals*. In many cases the change with time may be so slow as to permit broad classification as a stationary process. In others it will be necessary to partition the signals into short sections, short enough to consider the

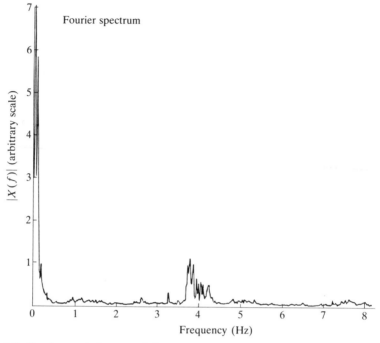

Fig. 1.2 Fourier spectral representation.

signal as almost stationary over the section, and to process each section separately.

It is convenient to expand the concept of a signal to include functions of independent variables such as distance, temperature, pressure, etc. For our purposes the functions $x(t)$, $x(s)$, x_i will be considered simply as single-dimensional functions having a linear base which could be time.

Multi-dimensional signals, that is, signals which refer to more than one variable, are expressed in a similar way. We have already met one of these as the non-stationary signal $x(t, f)$, which is a function of both time and frequency. A second common two-dimensional signal is that of a spatial description $x(m, n)$, or in discrete form $x_{i,j}$ which we meet in image processing, where (m, n) or (i, j) refer to Cartesian or other coordinates of the image area.

In general a multi-dimensional discrete signal may be expressed as $x_{i,k,\ldots,m}$. There exists considerable variation in methods of representation and storage of multi-dimensional sampled data [7]. Fundamental differences between handling these and single-dimensional data are associated with the quantity of data available when multi-dimensional information is considered. Some of the techniques developed for image transformation and processing will be considered later in this book.

1.2 Data acquisition

As mentioned earlier, signal representation for most physical processes is obtained in continuous analog form and requires conversion into discrete form for digital computer processing. This process involves sampling in the time domain, quantization in the amplitude domain, and coding the resulting information into digital form.

These three processes all impose limitations on the data obtained and so give rise to various sources of error which we should be aware of in later processing operations on the signal.

Sampling of the analog signal involves the selection of a series of narrow impulses or 'slices' of the signal, spaced at equal time intervals. A unique number is ascribed to each impulse and represents the mean amplitude of the sample taken over the area of the impulse. Ideally each sample should be taken over an infinitely short period of time but in a practical case it is necessary to estimate an averaged quantity over the sampling period. The length of time over which the data are averaged is known as the *aperture*. Aperture errors are minimized by the use of very fast multiplexers and analog-to-digital converter devices so that the time shift in the input–output sequences is negligible compared with the smallest signal period.

No matter how often we sample a continuous signal we are bound to lose some information in the process and this can give rise to a problem which is illustrated in Fig. 1.3. Here the same set of data points are able to describe a number of arbitrary signals which, after quantizing and coding, are indistinguishable to the digital computer. We say that these are 'aliased' signals. Thus for a signal $x(t)$ representing a waveform $\cos 2\pi f_0 t$, a set of aliased frequencies can be shown to exist which are related to the sampling interval h as:

$$f_0, \ 1/h - f_0, \ 1/h + f_0, \ 2/h - f_0, \ 2/h + f_0, \ \ldots, \ n/h \pm f_0. \quad (1.5)$$

The fundamental frequency f_0 is known as the *principal alias*. The range of frequencies below which this effect is not present extends from

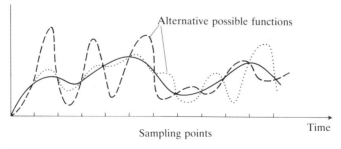

Fig. 1.3 Aliasing.

$f_0 = 0$ to $f_0 = f_N$. This maximum frequency f_N is the folding or *Nyquist frequency* and is a frequency limit–the Shannon limit to the sampled data, above which an unambiguous reconstruction of the signal is not possible [6]. Thus, given a signal having a bandwidth B Hz and containing no frequency components at and beyond a frequency f_N, then the lowest sampling frequency necessary to preserve the information contained in a sampled version of this signal is given as $f_s > 2B$, i.e. since $f_s = 1/h$, we have $B = 1/2h$. This is known as the *sampling theorem*. It follows that for a given frequency spectrum, the individual frequency components lying between $f = 0$ and $f = B$ can be separately examined, but if the signal contains components having frequencies $f > B$ they will not be distinguishable and will simply be added to the lower-frequency signals and contribute to the noise level associated with the signal.

The sampling theorem also applies to the time domain, so that if the signal of interest lies within a frequency band extending from 0 to B Hz then the minimum length of record necessary in order that we may recover this signal from the sampled data is

$$T \geqslant 1/2B \text{ s.} \tag{1.6}$$

From the preceding discussion we see that the sampling theorem consists of two parts, which may be restated as:

(1) Signals having a finite bandwidth up to and including B Hz can be completely described by specifying the values of the sampled signal series at particular instants of time separated by $1/2B$ s.

(2) If the signal is band-limited and contains no frequency greater than B Hz, it is theoretically possible to recover completely the original signal from a sampled version when the sampling interval is equal to or smaller than $1/2B$ s.

This concept of finite bandwidth is important. Consider a signal $x(t)$ to contain no frequencies higher than B Hz, i.e. $-B < f < B$. We can represent this in the frequency domain by its Fourier series $X(f)$:

$$X(f) = \frac{1}{2B} \sum_{n=-\infty}^{\infty} C_n \exp(-\mathrm{j}\pi n f/B), \tag{1.7}$$

where

$$C_n = \frac{1}{2B} \int_{-B}^{+B} X(f) \exp(\mathrm{j}\pi n f/B) \, \mathrm{d}f$$

and represents the complex Fourier amplitudes of the series. Now $X(f)$ fully defines the spectrum so that the equivalent time function $x(t)$ can be obtained from the inverse transform

$$x(t) = \int_{-B}^{+B} X(f) \exp(\mathrm{j}2\pi n f t) \, \mathrm{d}f. \tag{1.8}$$

If t is defined as $1/2B$, then eqn (1.8) becomes

$$x(n/2B)h = \int_{-B}^{+B} X(f)\exp(\mathrm{j}\pi nf/B)\,\mathrm{d}f$$
$$= 2BC_n. \tag{1.9}$$

From eqns (1.8) and (1.9) we see that if a band-limited function $X(f)$ is sampled at times $t = n/2B$ the original signal is completely recoverable from the sampled signal with no loss of information. Hence the sampling period must be $h \leqslant 1/2B$. From these considerations, it is seen to be essential that the signal be subjected to a low-pass filtering operation prior to digitization to ensure that all frequencies greater than $B = f_N$ are excluded.

This is necessary, not only to avoid the aliasing effects of the actual signal content, but also to reduce the contribution of higher-frequency noise components to the digitized data, which will otherwise be accepted as noise components falling within the Nyquist bandwidth B Hz. In practice this signal recovery is not perfect and the reason lies in the inadequacy of filter design. This may be seen if we consider the Fourier representation of a spectrum of limited bandwidth (Fig. 1.4) shown here as a two-sided spectrum. This necessarily includes an infinite series of spectra on either side of the original spectrum so that an anti-aliasing filter, having the characteristics shown dotted in Fig. 1.4, is required to precede the digital sampling process to permit full recovery of the original spectrum. Such perfect filter characteristics are unattainable and some distortion due to sampling is therefore inevitable.

Representation of a variable-amplitude series of discrete sample values as a limited series of discrete numbers is termed *quantization*. The process can only be an approximation since, while the original signal can assume an infinite number of states, the number of bits in a digital representation is limited. The numerical values of the quantized variable is represented by some form of binary code to permit entry into a digital computer or device.

In broad terms, quantization is a non-linear operation that is carried out whenever a physical quantity is represented numerically. The

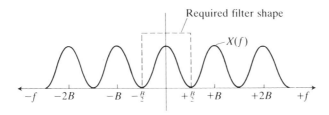

Fig. 1.4 Fourier representation of a limited bandwidth signal.

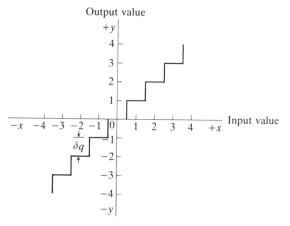

Fig. 1.5 Quantizer transfer characteristic.

resultant numerical value is given as an integer corresponding to the nearest whole number of units. This is expressed by the transfer characteristics of a quantizer shown in Fig. 1.5. An input value lying between the midpoint values of two consecutive unit values will produce an output value at the level corresponding to the higher of the two values.

The uncertainty or distortion introduced by this method of sample level determination can be considered as a form of noise additional to the signal [8]. This noise has the characteristics of *white noise,* i.e. it is uniformly distributed over the frequency band and is related to the number of quantization steps q contained in the dynamic range of the signal being quantized. The relationship can be shown to be (approximately)

$$S/N \simeq N/0.29 \qquad (1.10)$$

so that if, for example, the signal is quantized to at least $N = 256$ possible levels, then the *signal-to-noise ratio* (SNR) achieved will be approximately 1000, i.e. 60 dB in voltage ratio, [5] which is adequate for many purposes.

1.3 Digital filtering

An essential feature of most signal-processing software-package systems and hardware are routines for carrying out frequency filtering of the digital data. While in its most general sense a filter can be considered as a device used for selecting any particular frequency or set of frequencies, the term is usually confined to a system that transmits a certain range of frequencies, rejecting all others. Such frequency ranges are called pass-bands and stop-bands respectively, and the filters are categorized as low-pass, high-pass, and band-pass, referring to the range of frequencies transmitted in unattenuated form through the filter. A very brief look at the composition and characteristics of digital filters is contained in the following. For a more expansive treatment the reader is referred to the papers and textbooks listed at the end of this chapter.

A digital filter acts on a sampled version of the signal to be filtered and may be realized through discrete logical hardware elements or by suitable programming of a digital processor. The process of digital filtering represents an operation on a discrete series of input values x_i, such that the output series y_i is dependent on both the input series and a set of modifying coefficients defining the filter characteristics. This may be represented by means of a linear difference equation:

$$y_i = \sum_{k=0}^{P} b_k y_{i-k} + \sum_{k=0}^{M} a_k x_{i-k} \qquad (1.11)$$

$(i = 1, 2, \ldots, N)$, where P and M are positive integers and a_k and b_k are real constants.

1.3.1 IIR and FIR filters

When $M = 0$ the filter is auto-regressive, i.e. the output depends on the current input sample plus the weighted sum of past samples processed by the filter. This is known as the IIR—*infinite impulse response*—or recursive filter. The IIR filter will require relatively few terms to obtain an acceptable attenuation characteristic outside the pass-band but will have a non-zero phase characteristic, which may be unacceptable in some applications.

A number of design procedures have been developed [9] and, while some methods enable the design to be carried out completely in the frequency domain and hence use frequency transformation techniques, a more general approach is to apply the Z-transform and bilinear methods [5].

When $P = 0$ in eqn (1.11) the filter consists of the summation of the products of M weighting coefficients h_k, and only the present and past N samples of the signal waveform, i.e.

$$y_i = \sum_{k=0}^{M-1} h_k x_{i-k} \qquad (1.12)$$

$(i = 1, 2, \ldots, N)$. This is known as the FIR—*finite impulse response*—or non-recursive filter. This filter has excellent phase characteristics but requires a relatively large number of terms to obtain an acceptable attenuation characteristic. For this reason the direct evaluation through eqn (1.12) proves laborious and an alternative is to carry out FIR-filtering in an indirect fashion through domain transformation [5]. This involves a process of convolution, represented by ★, which may be considered for the moment as a form of summation for the product of data series. A more formal definition is given later in Section 1.4. With this method both the input data series x_i and the set of filter weights h_k are transformed into the frequency domain through the discrete Fourier transform (DFT) to give $X(\omega)$ and $H(\omega)$. The inverse discrete transform (IDFT) of the product of $X(\omega)$ and $H(\omega)$ gives the convolved series y_i, which itself represents the convolution process $x_i \star h_k$, and in effect realizes the FIR process of eqn (1.12) in another way. The process is expressed diagrammatically by eqn (1.13).

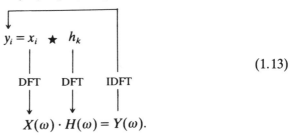

$$(1.13)$$

This method is particularly valuable with digital computation since it generally proves quicker to obtain a convolution of two data series by the product of their transforms and subsequent inverse transformation than by the direct product method of evaluating summations. In order to perform a filtering operation, however, it is necessary to compute the discrete Fourier transform of two sequences, namely the signal and filter weighting coefficient series, with the latter augmented to the same length as the signal. This lengthens the computation process and demands a large storage memory to hold the transformed coefficients for later multiplication. Two methods of overcoming these difficulties are the *select-save* and *overlap-add* methods [10, 11]. Both rely on filtering subsets of the input sampled signal using this frequency-domain approach and then recombining the sub-sequences obtained before going on to carry out similar filtering operations on the remaining subsets. Further by taking advantage of certain characteristics of the convolution properties of the real and imaginary filtering segments it is possible to process two segments in a single iteration, thus halving the number of iterations in the procedure as well as reducing the storage-memory requirements. Several detailed implementations of these FIR methods are described in the literature [9, 12].

1.3.2 Spectral windows

There is a difficulty arising from using a Fourier transformation as a route to the discrete digital filter. As pointed out earlier the ideal filter

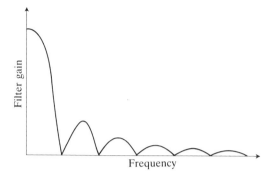

Fig. 1.6 Frequency response of a truncated digital low-pass filter.

characteristic is one in which infinite attenuation is given to a signal lying outside the filter pass-band and zero attenuation is given within the pass-band. Unfortunately such characteristics, or even a near approach to them, would result in a Fourier series with a very large (tending towards infinite) number of coefficients requiring a correspondingly large number of filter coefficients. In order to implement such a filter it is necessary to truncate in some way the Fourier series so obtained. This in turn leads to an oscillatory form of the filter frequency response in the region near a discontinuity, i.e. the point at which the response changes from zero to maximum attenuation or vice versa (Fig. 1.6), and is called the 'Gibbs phenomenon'. Some alternative way of reducing the length of the Fourier series for a filter representation must then be found. First however, let us look at this process of truncation in a little more detail.

It can be shown [9] that the truncation of a discrete Fourier series of n values corresponds to a multiplication (actually a convolution) of an infinite Fourier series with a function w_n given by

$$w_n \begin{cases} =1 & \text{if } -\tfrac{1}{2}(N-1) \leqslant n \leqslant \tfrac{1}{2}(N-1), \\ =0 & \text{otherwise} \end{cases} \tag{1.14}$$

$(n = 0, 1, 2, \ldots, N-1)$ where N is the number of filter coefficients.

This is a time function representing the discontinuity and is known as a *window function*; the 'window' through which we 'see' a section of an infinite Fourier series (Fig. 1.7(a)). As with any other function we can express this as having a frequency response, through the use of Fourier transformation as,

$$w_n(\omega) = (\sin(N\omega/2))/(\omega/2) \tag{1.15}$$

This is a real-valued function of ω and is plotted in Fig. 1.7(b). From this we note that the oscillatory response of the truncation process consists of a main lobe and several side-lobes of considerably smaller areas. As the number of terms we use N increases the main lobe becomes

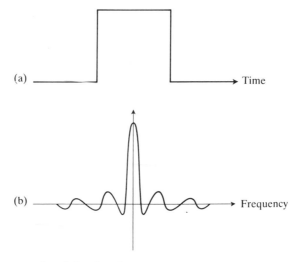

Fig. 1.7 A rectangular window function
(a) in the time domain (b) in the frequency domain.

higher and narrower so that as $N \rightarrow \infty$ the oscillatory response diminishes and eventually vanishes altogether.

The concept of separation of the filter frequency characteristic into the product of an ideal filter response and a window function w_n with which it is convolved, provides a method of modifying this oscillatory distortion error in the FIR filter and, as we shall see later, a similar distortion problem encountered in spectral density evaluation.

We do this by convolving the truncated filter signal with a further modifying window. The characteristics of this added window are chosen to minimize the oscillatory response shown in Fig. 1.6. This means in general terms that the window frequency response needs to exhibit a small width of main lobe and side lobes that decrease rapidly in value. The window function represents a compromise between conflicting design requirements so that a number of alternative window functions have become available for the design of FIR digital filters.

Some of the most common windows are:

(a) the Hanning window

$$\begin{aligned} w_n &= 0.5 + 0.5 \cos(n\pi/N) \quad &\text{for } |n| < N \\ &= 0 &\text{for } |n| \geq N. \end{aligned} \tag{1.16}$$

(b) the Hamming window

$$\begin{aligned} w_n &= 0.54 + 0.46 \cos(2\pi n)/N \quad &\text{for } |n| < N \\ &= 0 &\text{for } |n| \geq N. \end{aligned} \tag{1.17}$$

(c) the Blackman window

$$w_n = 0.42 + 0.5 \cos(\pi n)/N + 0.08 \cos(2\pi n)/N \quad \text{for } |n| < N$$
$$= 0 \qquad\qquad\qquad\qquad\qquad\qquad\qquad \text{for } |n| \geqslant N. \tag{1.18}$$

(d) the Bartlett (or triangular) window

$$w_n = [1 - 2\,|n - N/2|/N] \quad \text{for } |n| < N$$
$$= 0 \qquad\qquad\qquad\qquad \text{for } |n| \leqslant N. \tag{1.19}$$

Many other windows are described elsewhere [9, 13] and a choice needs to be made between the complexity of implementation and the lowest accuracy in spectral representation necessary for the application. The Hanning window is a common solution that is fairly easy to apply, and its implementation will be considered later.

1.3.3 Wiener filtering

Another approach to the filtering problem is to consider FIR filtering in terms of classic Wiener filtering [14]. This applies matrix methods to the solution of the filtering problem in a way very suited to the deployment of a number of alternative orthogonal transforms so that the method is not confined to the Fourier transform alone, as with the methods discussed above.

Figure 1.8 is a block diagram of a single-dimensional generalized Wiener filtering system applied to SNR enhancement. A signal x_i is assumed to consist of additive zero-mean signal s_i and noise n_i components, which are assumed to be uncorrelated with each other. A transformation operation using a transformation matrix, A (considered in the next chapter) is carried out on x_i to yield

$$X_k = A \cdot x_i = A \cdot s_i + A \cdot n_i$$
$$= S_k + N_k. \tag{1.20}$$

The resultant transformed values X_k are then multiplied by an $N \times N$ filter matrix G and inversely transformed to produce a filtered output

$$y_i = A^{-1} \cdot G \cdot A \cdot x_i. \tag{1.21}$$

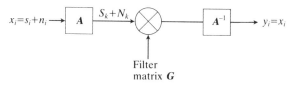

Fig. 1.8 A single-dimensional Wiener filtering system.

If **G** is chosen correctly then the required filter output will consist of the signal component s_i plus a much reduced noise component n_i. The filter design centres on the choice of filter matrix G, which may be obtained through a minimization of the mean-square error (MSE) between the required signal s_i and its estimated value $s_i = y_i$ [15]. In many cases the filter matrix **G** can be reduced to a diagonal vector (see later in Chapter 4), reducing the calculation to a single-dimensional matrix and thus minimizing the amount of computation required.

An important feature of the Wiener filtering method is that it can be applied using any orthogonal transform without affecting the MSE of the process. This means that one is free to choose a transform that minimizes the computational processes required so that good results may be obtained in some cases using non-sinusoidal transformations, which are nearly always faster to compute than those based on the Fourier transform.

1.4 Correlation and convolution

Much of the previous discussion has been concerned with the characteristics and manipulation of individual signals. We may however wish to compare signals one with another, and although it may be useful to compare single characteristics such as their mean values or peak amplitudes, it is more meaningful to obtain a quantitative measure for the shared property of the two signals.

We find this in *correlation* and the derivation of the *correlation coefficient*. Given two digital series, x_i and y_i we attempt to find a series of average product values between them for differing values of their relative time delay. This apparently roundabout process is necessary to take into account a measure of similarity that may be present when one series is delayed in time relative to the other.

The result of this time-delayed set of average products describes the *cross-correlation function*, which is given by,

$$R_{xy}(\tau) = 1/(N - \tau) \sum_{i=1}^{N-1} x_i y_{i+\tau} \qquad (1.22)$$

$(\tau = 0, 1, \ldots, m)$, where m is the total number of correlation delay or lag values we wish to use and N is the number of data values.

This equation gives an unbiased estimate of the ensemble average for a finite time period and shows that the summation period will decrease as the correlation time lag τ is made larger. We would therefore expect the reliability of this estimate to be reduced as the lag approaches the record length N. However for $\tau < N$ we can with little loss of accuracy replace eqn (1.22) by

$$R_{xy}(\tau) = 1/N \sum_{i=1}^{N-\tau} x_i y_i + \tau.$$ (1.23)

The direct evaluation of eqn (1.23) demands much computing time for large N values and, as with the convolution approach to digital filtering described earlier, it is faster to obtain $R_{xy}(\tau)$ through eqn (1.12) by carrying out two forward Fourier transformations, one inverse transformation and one multiplication per data sample pair. Some care has to be taken in applying fast transform methods, however, to avoid circular correlation distortion, which is an inevitable result of representing a finite signal with a finite discrete frequency series.

1.4.1   Circular correlation

Circular correlation results from the correlation of periodic signals of finite period N, which we get when the product of two transformed signals is taken (eqn (1.13)). The sequences to be correlated repeat in a circular fashion so that for a sequence $x_0, x_1, x_2, \ldots, x_{N-1}$ the next value in the sequence will be x_0, etc. This is illustrated in Fig. 1.9(a). Consider the sequence x_n to be convolved with another sequence $y_0, y_1, y_2, \ldots, y_{M-1}$ shown in Fig. 1.9(b). The convolution is effectively the sum of products of terms $x_k y_k$ that are shown directly underneath one another in the diagram. As the sequence y_n is shifted past x_n by the incremental lag value τ, the correlated product-sums are evaluated. The process does not stop when $x_n = x_{N-1}$ is reached, since the next value to be included in the correlation calculation is x_0, followed by x_1 etc. This

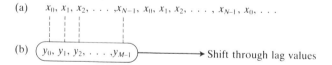

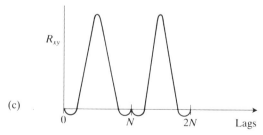

Fig. 1.9   Effects of circular correlation.

new sum is added to the first correlation product-sum obtained and thus introduces an unwanted distortion of the result. The way to avoid this is to add N zeros (equal to the length of the signal) to both the signals before transformation, which results in the correlogram shown in Fig. 1.9(c). Here the repeated but imaged signal can then easily be truncated to obtain the first half of the correlated signal required.

In the above discussion the terms convolution and correlation have both been used. Convolution is similar to the correlation process but with a reversal of one of the time functions, i.e.

$$C_{xy}(\tau) = 1/(N - \tau) \sum_{i=1}^{N-1} x_i y_{\tau - i} \qquad (1.24)$$

(compare eqns (1.22) and (1.24)).

The convolution of x and y is often written in the form $x \star y$ instead of as shown in eqn (1.24). This form will be used frequently in the following.

Cross-correlation gives an indication of the joint properties shared by the two sampled signals, x_i and y_i. Since the similarities between the two signals may become apparent only after a time delay, and in general we do not know what this time delay is, we are forced to correlate $y_{i+\tau}$ with x_i over a number of possible values for τ. The result is a *cross-correlogram* which conveys information concerning phase differences between the two signals as well as their similarity. An example is given in Fig. 1.10 for two similar but phase shifted signals.

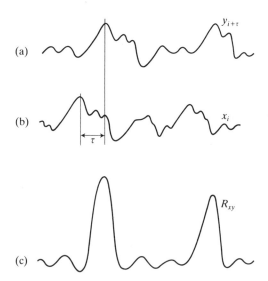

Fig. 1.10   Cross-correlogram of two similar signals.

This phase information is absent with an *auto-correlation*, which gives the relation between values of the same process x_i but compared at different times separated by a time lag τ. This is expressed for a digital series x_i as

$$R_{xx}(\tau) = 1/N \sum_{i=1}^{N-1} x_i x_{i+\tau} \tag{1.25}$$

$(\tau = 0, 1, \ldots, M)$.

1.4.2 Correlation with real signals

The particular importance of the auto-correlation function lies in the fact that it tells us something about the frequency domain behaviour of x_i. Some examples will make this clear.

Figure 1.11 shows the auto-correlation of a wide-band noise signal. The auto-correlation is nearly zero except for a sharp peak at $\tau = 0$, showing that x_i at different times are nearly unrelated with each other and that only a small time shift destroys the similarity between x_i and $x_{i+\tau}$. In contrast, a sinusoidal signal as shown in Fig. 1.12 has a correlation function which persists over a range of time lags, showing a cyclic similarity between different values.

Correlation methods are much used in the basic problem of improving the SNR of a signal obscured by noise, which itself may occupy the same frequency bandwidth [16]. This is not easy to achieve through conventional filtering methods which rely on some form of frequency separation between the wanted and unwanted signal. We can however take advantage of the phase characteristics of the correlation process to modify the signal we wish to detect so that it can more readily be identified in this way.

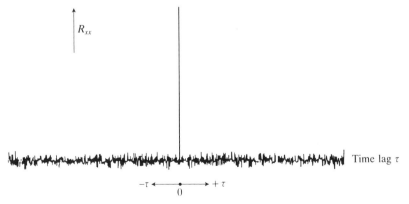

Fig. 1.11 The auto-correlation of a wide-band noise signal.

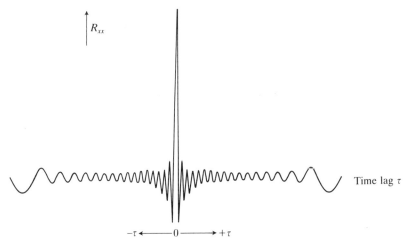

Fig. 1.12   The auto-correlation of a sinusoidal function.

Consider for example the needs of radar signal processing where the direction and position of a distant object are determined by the reception of reflected signals which are usually obscured by noise. The ability to detect the return signal reflected by a target can be shown to be directly proportional to the product of the average signal power and the signal duration. Practical limitations in transmitter design control the peak value of the signal and it becomes necessary to obtain the required range by increasing the signal duration. Unfortunately the reduction in band-width which this entails also leads to a reduction in signal resolution, i.e. the ability to distinguish between echoes derived from closely spaced objects. A solution to this resolution/range problem lies in the replace-ment of the long-duration, narrow-band pulse by a *swept-frequency signal* having both long duration and wide bandwidth. The waveform used is shown in Fig. 1.13(a); in another context (Chapter 6) this is called a *chirp signal.*

Detection of the swept-frequency echo signal and the separation of this from its associated noise is carried out using cross-correlation of the signal with a reference waveform derived from the transmitter output. Ideally the signal and the reference swept-frequency waveform would be identical, apart from the time delay between them. Hence the correlation process will compress the energy dispersed throughout the long swept frequency signal into a short well-defined pulse of consequently enhanced average power. This is shown in Fig. 1.13(b) and illustrates the value of this technique in raising the correlated pulse to a value well clear of the noise level. A *threshold detection* can then be carried out to give a precise location of the time position of the returned pulse.

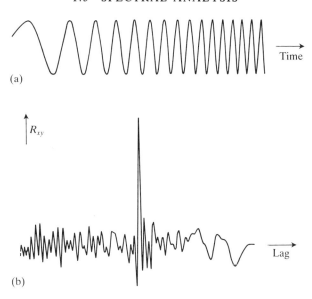

(a)

(b)

Fig. 1.13 Correlation detection of a swept frequency signal.

1.5 Spectral analysis

Determination of spectral distribution is a major application of transformation theory. Although the spectral decomposition of a signal can be obtained from the auto-correllogram a more useful description is obtained by a direct plot of its spectral content. Fourier series analysis is inadmissible for random signals which do not necessarily have harmonically related components. Instead a measure of the relative amplitude of the frequency components is obtained via the Fourier transform,

$$X(f) = \int_{-\infty}^{\infty} x(t)\exp(-\mathrm{j}\omega t)\,\mathrm{d}t \qquad (1.26)$$

($\omega = 2\pi f$) or in discrete form as

$$X_k = 1/N \sum_{i=0}^{N-1} x_i \exp(-\mathrm{j}2\pi ik/N). \qquad (1.27)$$

Due to the random nature of most physical signals we need an average spectral measure that is relatively stable and meaningful to interpret. We find this in the energy content distributed over the frequency bandwidth generally referred to as the *power spectral density* (PSD) of the signal. For a sampled signal this is given as

$$P_k = 1/N \left| \sum_{i=1}^{N-1} x_i w_i \exp(-j2\pi ki) \right|^2 \qquad (1.28)$$

where w_i is a window function inserted to minimize a fundamental difficulty of Fourier analysis mentioned earlier, namely the identification of harmonic components whose periods are not integer factors of the analysis interval. The effect of this window operation is to carry out averaging operations on the spectrum to reduce the variance of the spectral estimate. The spectrum we actually 'see' through the window is never the true spectrum but an estimate whose accuracy is dependant on several factors, including the choice of the window itself.

There are in fact three well-known methods of arriving at the PSD [17]. One is the direct evaluation of P_k through eqn (1.28), which is termed the *Periodogram* method. The second is via the autocorrelation function and is known as the *Blackman–Tukey* method. A third relates to the use of the fast Fourier transform in a *segmentation method*. In this latter method the signal is divided into M segments, a Fourier transformation is carried out over each segment, the coefficients are squared, and an average is taken over all the segments. Smoothing of the spectrum obtained is thus inherent in this method. Its advantage lies in computational simplicity, since we do not have to first calculate an autocorrelation series. It does require some time to carry out the averaging process but has the advantage of requiring much less computing memory [18].

These may be termed the 'older' methods of PSD estimation. They are still in wide use for very many applications.

In the last decade some alternative procedures have been developed that are essentially data modelling methods based on statistical approaches. The goal is to create a spectrum for the signal based on a set of characteristics or parameters derived from statistical analysis of the signal [19]. What is actually determined is not the spectrum of the signal itself but a synthetic spectrum constructed out of these parameters. Four variants of these methods may be identified under the generic description of *parametric methods*. They are linear prediction, autoregression, maximum entropy, and maximum likelihood. They are general analysis techniques applicable not only to PSD evaluation but also to study properties of the data in the time domain. They are particularly valuable in pattern recognition and non-stationary analysis.

In the following pages the various techniques for PSD estimation, both 'old' and 'new', will be considered in a little more detail, beginning first with the Periodogram.

1.5.1 The Periodogram

An acceptable procedure to derive the discrete PSD is to separate the process into a calculation of the 'raw' spectral estimate,

$$P_k = 1/N \, |X_k|^2 \qquad (1.29)$$

where X_k is the Fourier transform of the time series x_i, and to process P_k with a windowing operation to improve the accuracy of the estimate. X_k is obtained in terms of the real [Re.] and imaginary [Im.] coefficients so that the estimate is more properly shown as,

$$P_k = 1/N \, |\text{Re}^2(X_k) + \text{Im}^2(X_k)|. \tag{1.30}$$

Of the smoothing window routines available the Hanning window is probably the easiest to apply. Equation (1.15) gave an expression for a Hanning *lag window* as a weighting function to be convolved with the signal series. This may be combined with the raw spectral estimate of eqn (1.29) to yield an expression for a smoothed spectral estimate, P_k [20]. Here we find the running average of three sets of P_k terms based on eqn (1.29) in which the first set uses the $(k-1)$th coefficient of X_k weighted by a constant, 0.25; the second set uses the current coefficient k weighted by 0.5; and the third set uses the $(k+1)$th coefficient weighted by 0.25. As k is advanced to determine each averaged coefficient value for P_k a smoothed discrete value for this is realized based effectively on the average of three adjacent values. The overlapping of these sets of values as k advances produces a running average set of coefficients for P_k, i.e.

$$\hat{P}_k = 1/N[0.25\,|X_{k-1}|^2 + 0.5\,|X_k|^2 + 0.25\,|X_{k+1}|^2] \tag{1.31}$$

$(k = 1, 2, \ldots, N)$. The smoothed estimate value $\hat{P}_k$ is usually quite adequate for most purposes. Other methods of smoothing are described extensively in the literature (see refs. in [17]).

1.5.2 The Blackman–Tukey method

This is an indirect method in which the auto-correlation of the signal is first calculated and advantage taken of the relationship between the auto-correlation function and the PSD, which is obtained through the *Wiener–Khintchine theorem* as

$$P_k = 1/M \sum_{i=0}^{M-1} R_{xx}(i)\exp(-\text{j}2\pi ki) \tag{1.32}$$

$(i = 0, 1, \ldots, M)$. i is here taken as the number of correlation lags, which are generally limited to about 10 per cent of the available number of samples, i.e. $M = N/10$. Again smoothing through a windowing technique is applied to this method of spectral estimation [20].

The auto-correlation approach has the advantage of producing the auto-correlogram as an intermediate result and generally requires less computer memory than the Periodogram method.

1.5.3   *The segmentation method*

The sampled data is divided into segments of length T, which relate to the lowest frequency for which the spectral estimation is of interest. If the number of samples included in T is very small, say less than 64, there is no advantage gained from using the fast transform techniques described later in Chapters 4 and 5. For each segment the real and imaginary parts of the discrete Fourier transform are determined and the sum of their squares added as given in eqn (1.30). The raw spectrum is then convolved with a window function, w_n and the average of the set of 'windowed' spectra determined to arrive at the smoothed PSD.

1.5.4   *Data modelling*

Thus far spectral analysis has been considered as a process of estimating the power spectrum from a finite duration signal sample. This presents a compromise between accuracy and spectral resolution, which for short samples of data is very difficult to resolve. During the last decade a new set of analysis methods, known as the parametric methods, have been developed, which are particularly attractive for making high-resolution spectral estimates when the data record is short.

Using these methods a set of parameters are obtained relating to the signal, which are used to control a model synthesis of a spectrum where the input to the process is a random signal exhibiting a uniform spectrum. This is, in effect, the definition of 'white noise'. The structure of the signal generator model used for synthesizing the signal out of this white noise (which must, by definition, contain the signal to be synthesized) determines the type of parametric process applied.

Several models have been used. These were stated earlier as the auto-regressive (AR), auto-regressive moving average (ARMA), linear prediction (LP), maximum entropy (ME), and maximum likelihood (ML) methods [21]. None of these methods use window functions and all are of the data adaptive type; that is, the methods adjust themselves to be least disturbed by power at other frequencies than those contained in the signal.

A brief look at one of these, the maximum entropy method (MEM), will indicate the type of thinking behind their development.

If we apply one of the earlier methods of spectral estimation to a process that has only a few sharp and closely spaced peaks we are likely to find that the peaks will be poorly resolved, as the use of a smoothing window tends to merge fine details together. What we are up against is the basic weakness of the Fourier transform method in that it assumes that any frequency is potentially of equal importance. When this is not true, the method cannot perform well. If we relate this to the Blackman–

Tukey method for example, we find a conflict between good resolution, which requires the auto-correlation function to be known or estimated out to large time lag values, and good accuracy, which requires auto-correlation functions to be estimated only out to small time lag values.

The maximum entropy method behaves differently because it tends to reproduce the most prominent feature of the power spectrum. Less prominent structures are reduced and may be lost. Some care is therefore necessary when choosing this method since it must be applicable to the particular spectrum being estimated [21].

Van den Bos [22] observed that the MEM is like a least-squares fitting of an auto-regressive (feedback) model to the available signal data. Errors are modelled by white noise using a configuration such as the one shown in Fig. 1.14. The white noise z_n is used as input to a filter, modified by a gain coefficient G, and combined with the filter output y_n, i.e.

$$x_n = z_n G \sum_{k=0}^{P} b_k x_{n-k} \tag{1.33}$$

where G is a gain factor and P is the number of filter delays.

The filter coefficients are determined from the auto-correlation coefficients of the actual signal x_n, although the signal itself is not input into the model regressive system. The square of the inverse transformation of the parameters of the filters b_0 to b_P will give an estimate of the power spectrum [23], i.e.

$$P_n = 1/P \left| \sum_{k=0}^{P} b_k \exp(j2\pi fkD) \right|^2 \tag{1.34}$$

where D is the filter unit lag.

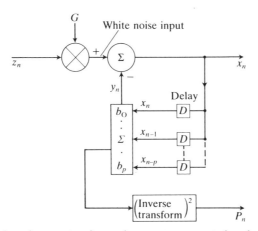

Fig. 1.14 Derivation of parameters for maximum entropy spectral analysis.

This indirect method allows us a choice of filter lags which is not bounded by the limited duration of the signal and in effect avoids the Fourier disadvantage of *a priori* selection of uniformly spaced model frequencies. Many implementations of this and the other parametric methods are described in the literature. An excellent review of modern spectral analysis is given by Kay and Marple [24] and selected reprint papers available in Childers [25]. See also [22, 23, 26, 27].

1.5.5 Cepstrum techniques

Finally, a different form of spectral evaluation is considered that has close similarities with the method of auto-correlation, which we saw earlier contains spectral information within its characteristics. This is known as the *cepstrum* (a corruption of the term spectrum) [28]. Whereas auto-correlation may be defined as the Fourier transform of the power spectrum, the cepstrum is defined as the Fourier transform of the logarithm of the power spectrum,

$$K_n = 1/N \sum_{k=0}^{N-1} (\log_e |X_k|^2) \exp(2\pi kn/N). \tag{1.36}$$

The principle use for the cepstrum lies in the unambiguous indication of the rate of repetition for the component parts of periodic signals that have a very non-linear form. Such signals are rich in harmonically-related terms and an auto-correlogram would present a complicated indication in the time domain due to the profusion of extracted periodicities. In effect what the cepstrum method does is to carry out Fourier transformation twice, once on the composite harmonic signal $x(t)$ to produce a spectrum value $X(f)$, and then on $\log_e X(f)$ to separate out the products of the required signal and its harmonics contained in $x(t)$. It has been used to separate out a particular structural resonance in vibration machinery [29] and in image processing applications to resolve the problems of motion and defocus blur or poorly represented pictures [30] and will be referred to later in connection with speech and sonar processing.

1.6 Two-dimensional processing

Essentially we have been considering only single-dimensional signal processing methods in this introductory chapter. Representation of a two-dimensional signal or image implies that sampling must take place at a finite number of discrete spatial points. Not all the information in the image is being used in this representation and it is more usual to take an average value of light intensity over a sub-area of the picture known as a *pixel* or picture element. The accuracy of digital representation is dependent on the number of pixels comprising the complete image (and of course on the number of discrete grey level values used for image representation).

Image processing is necessarily more complex than single-dimensional processing. The difficulties are associated first with the size of the database. It is important to choose methods that reduce the scale of the problem, generally by increasing the speed of processing. This is reflected in the selection of processing algorithms, particularly those used for domain transformation. There are additional problems such as incipient instability in IIR filters and additional information introduced by colour imaging. These topics however fall outside the scope of this book, and only the subject of two-dimensional transformation will be considered. For the most part this will be expressed simply in terms of ways of representing multi-dimensional transformation as a sequential set of single-dimensional transforms, so that the transformation process itself will be inseparable from the general treatment given here of single-dimensional transformation.

References

1. Gold, B. and Rader, C. M. *Digital processing of signals*. McGraw-Hill, New York, (1969).
2. Oppenheim, A. V. and Schafer, R. W. *Digital signal processing*. Prentice-Hall, Englewood Cliffs (1975).
3. Rabiner, L. R. and Gold, B. *Theory and application of digital signal processing*. Prentice-Hall, Englewood Cliffs (1975).
4. Blackman, R. B. and Tukey, J. W. *The measurement of power spectra*. Dover, New York (1959).
5. Beauchamp, K. G. and Yuen, C. K. *Digital methods for signal analysis*. George Allen and Unwin, London (1979).
6. Shannon, C. E. A mathematical theory of communication. *Bell Syst. tech. J.* **27,** 623–56 (1948).
7. Huang, T. S. *Picture processing and digital filters*. Springer, Heidelburg (1975).
8. Widrow, B. Statistical analysis of amplitude quantised sampled data systems. *IEEE Trans. Appl. Ind.* **ID-52,** 555 (1961).
9. Cappellini, V. Constantinides, A. G., and Emiliani, P. *Digital filters and their applications*. Academic Press, London (1978).
10. Stockham, T. G. High-speed convolution and correlation. *AFIPS Proc. Spring Joint Comput. Conf.* **28,** 229–33 (1966).
11. Helms, H. D. Fast Fourier transform method of computing difference equations and simulating filters. *IEEE Trans. Audio Electroacoust.* **AU-15,** 89–90. (1967).
12. Capellini, V. and Constantinides, A. G., ed. *Digital signal processing*. Academic Press, London (1980).
13. Kaiser, J. F. Digital filters. In *System analysis of digital computers* (ed. F. F. Kuo and J. F. Kaiser). Wiley, New York (1966).
14. Wiener, N. *The extrapolation and smoothing of stationary time series*. MIT Press, Cambridge, MA (1949).

15. Pratt, W. K. Generalized Wiener filtering computational techniques. *IEEE Trans. Comput.* **C-21** (7), 636–41 (1972).
16. Lee, Y. W., Cheatham, T. P., and Wiesner, J. B. Application of correlation analysis to the detection of periodic signals in noise. *Proc. IRE* **38,** 1165–71 (1950).
17. Yuen, C. K. and Fraser, D. *Digital spectral analysis.* Pitman, San Francisco (1979).
18. Welch, P. D. The use of fast Fourier transform for the estimation of power spectra; a method based on time-averaging over short modified periodograms. *IEEE Trans. Audio Electroacoust.* **AU-15,** 70–3, (1967).
19. Childers, D. G., ed. *Modern spectrum analysis.* IEEE Press, Wiley, New York (1978).
20. Bendat, J. S. and Piersol, A. G. *Measurement and analysis of random data.* Wiley, New York (1966).
21. Frost, O. L. Power spectrum estimation. NATO ASI on Signal Processing, Portovenere, Italy (1976).
22. Van den Bos, A. Alternative interpretation of maximum entropy. *IEEE Trans. inf. Theory* **IT-17,** 493–4 (1971).
23. Burg, J. P. *Maximum entropy spectral analysis.* Ph.D. thesis, Stanford University, Palo Alto, CA (1975).
24. Kay, S. M. and Marple, S. L. Spectral analysis—a modern perspective. *Proc. IEEE* **69,** 1380–419 (1981).
25. Childers, D. G., ed. *Modern spectral analysis.* IEEE Press, Wiley, New York (1978).
26. Lacoss, R. T. Data adaptive spectral analysis methods. *Geophysics* **36**(4), 661–75 (1971).
27. Ulrych, T. J. and Bishop, T. N. Maximum entropy spectral analysis and auto-regressive decomposition. *Rev. Geophys. Space Phys.* **13,** 183–200. (1975).
28. Bogert, B. P., Healey, M. J. R., and Tukey, J. W. The quefrequency analysis of time series for echoes, cepstrum, pseudo-autocovariance, cross-cepstrum and saphe cracking. In *Proceedings of the symposium on time series analysis* (ed. M. Rosenblatt). Wiley, New York (1963).
29. Wellstead, P. E. Self-tuning digital control systems. In *Computer control of industrial processes.* Peter Peregrinus, IEE, London (1982).
30. Cannon, M. Blind deconvolution of spatially invariant image blurrs with phase. *IEEE Trans. acoust. Speech signal Process.* **ASSP-24**(1), 58–63. (1976).

2

ORTHOGONALITY AND DOMAIN TRANSFORMATION

2.1 Orthogonality

It is often important to represent a signal waveform by means of a series of formalized waveforms to bring some order to the process of analysis. We can use as many terms as we like—the more we use the greater the accuracy in our realization. The great advantage of such a simulation is that we are then in a position of being able to analyse a signal with some knowledge of its constituent parts. To succeed we need to choose a set of continuous functions, $f(t) = f_0, f_1, f_2, \ldots$ that are mutually *orthogonal* over the interval t_0 to a maximum time interval T. A criterion for this is when

$$\int_{t=t_0}^{T} f_m(t)f_n(t)\,dt = c \quad \text{if } m = n$$

$$= 0 \quad \text{if } m \neq n \tag{2.1}$$

where c is a constant.

If c is arranged to be one, the function set is said to be *orthonormal* [1]. The significance of this important relationship is seen if we look first at the case of a periodic sinusoidal function such as the series given in eqn (1.1) and presented here in a more general form as,

$$x(t) = a_0 + a_1 \cos(\omega_0 t) + b_1 \sin(\omega_0 t) + \ldots + a_n \cos(n\omega_0 t) + b_n \sin(n\omega_0 t) \tag{2.2}$$

for n terms, which we can write as

$$x(t) = a_0 + \sum_{k=1}^{n} a_k \cos(k\omega_0 t) + \sum_{k=1}^{n} b_k \sin(k\omega_0 t). \tag{2.3}$$

Equation (2.3) is known as the *Fourier series* for the function, $x(t)$ and a_0, a_k, and b_k the *Fourier coefficients*.

It will be obvious that the area under a sinusoidal or cosinusoidal waveform for a complete period $T = 1/f_0$ must be zero, i.e.

$$1/T \int_{0}^{T} \sin(n\omega_0 t)\,dt = -1/(n\omega_0 T)\,|\cos(n\omega_0 t)|_0^T$$

$$= -1/(n\omega_0 T)[\cos(2\pi n) - \cos 0] = 0 \tag{2.4}$$

and similarly for

$$1/T \int_0^T \cos(n\omega_0 t)\, dt$$

where n is an integer.

Also writing

$$\cos(m\omega_0 t)\cos(n\omega_0 t) = 1/2(\cos(m+n)\omega_0 t + \cos(m-n)\omega_0 t) \quad (2.5)$$

we see that if m and n are unequal integers, then

$$1/T \int_0^T \cos(n\omega_0 t)\cos(n\omega_0 t)\, dt = 0 \quad (2.6)$$

similarly

$$1/T \int_0^T \sin(n\omega_0 t)\sin(n\omega_0 t)\, dt = 0 \quad (2.7)$$

and

$$1/T \int_0^T \sin(n\omega_0 t)\cos(n\omega_0 t)\, dt = 0. \quad (2.8)$$

The only integrals of this type which have a finite value are

$$1/T \int_0^T \sin^2(n\omega_0 t)\, dt = 1/T \int_0^T \cos^2(n\omega_0 t)\, dt = 1/2. \quad (2.9)$$

Thus a finite value is obtained for a weighted average of the product of sine and sine, or cosine and cosine if their frequencies and phase shifts are identical, and a zero result if they are not. Also the weighted average of the product of sine and cosine will be zero irrespective of their frequencies. These are the orthogonal features shown in eqn (2.1) for two sine/cosine functions and are the key to many analysis techniques used in signal processing, the most important of which is spectral analysis.

Using these results it can be shown that the Fourier coefficients a_0, a_k, and b_k, which represent the amplitudes of the harmonics in the series $x(t)$, are obtained as

$$a_0 = 2/T \int_0^T x(t)\, dt \quad (2.10)$$

$$a_k = 2/T \int_0^T x(t)\cos(k\omega_0 t)\, dt \quad (2.11)$$

$$b_k = 2/T \int_0^T x(t)\sin(k\omega_0 t)\, dt. \quad (2.12)$$

A plot of these coefficients gives information about the frequency content of the function and is known as the *Fourier spectrum* (Fig. 2.1).

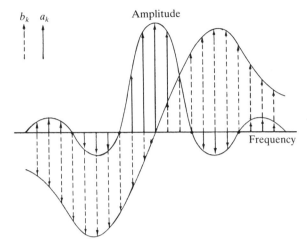

Fig. 2.1 Fourier spectral coefficients.

It is readily apparent that we can reconstitute the original function, $x(t)$ from a set of such harmonic sine and cosine waveforms providing we know the amplitudes of the constituent coefficients a_k and b_k. We do not even need to know all of these and an acceptable simulation can often be made from a limited number of coefficient values. Hence the representation of a waveform in the spectral domain by means of a limited number of its spectral values can be quite accurate and more economical in data storage than its time domain representation. An exact relationship between the two domains of representation can be obtained through the distribution of power amongst the various coefficients. This may be expressed through *Parseval's theorem* [2] as

$$1/T \int x^2(t)\, \mathrm{d}t = a_0^2 + 1/2 \sum_{n=1}^{\infty} (a_n^2 + b_n^2) \qquad (2.13)$$

which equates the 'energy' in the signal with the 'energy' in its spectrum.

2.1.1 Complex representation of Fourier series

Complex representation of Fourier series and the Fourier integral simplifies the notation and allows later development of the Fourier transform. Referring to eqn (2.3) we can expand the term

$$a_k \cos(k\omega_0 t) + b_k \sin(k\omega_0 t)$$

using the complex relationship,

$$\cos(k\omega_0 t) = 1/2[\exp(\mathrm{j}k\omega_0 t) + \exp(-\mathrm{j}k\omega_0 t)] \qquad (2.14)$$

and

$$j \sin(k\omega_0 t) = 1/2[\exp(jk\omega_0 t) - \exp(-jk\omega_0 t)] \qquad (2.15)$$

where $j = \sqrt{-1}$. Thus

$$
\begin{aligned}
(a_k &\cos(k\omega_0 t) + b_k \sin(k\omega_0 t) \\
&= a_k/2[\exp(jk\omega_0 t) + \exp(-jk\omega_0 t)] \\
&\quad + b_k/2[\exp(jk\omega_0 t) - \exp(-jk\omega_0 t)] \\
&= A_k \exp(jk\omega_0 t) + B_k \exp(-jk\omega_0 t)]
\end{aligned}
\qquad (2.16)
$$

where

$$A_k = (a_k - jb_k)/2 \quad \text{and} \quad B_k = (a_k + jb_k)/2. \qquad (2.17)$$

A_k and B_k represent the *complex conjugate amplitude coefficients* of the Fourier series and can be expanded by eqns (2.10) and (2.11) to find the complex Fourier series. For convenience the variable of integration in these equations will be changed to p in the following. Thus

$$
\begin{aligned}
A_k &= 1/T \int_0^T x(p)[\cos(k\omega_0 p) - j \sin(k\omega_0 p)]\, dp \\
&= 1/T \int_0^T x(p)\exp(-jk\omega_0 p)\, dp
\end{aligned}
\qquad (2.18)
$$

and similarly

$$B_k = 1/T \int_0^T x(p)\exp(jk\omega_0 p)\, dp. \qquad (2.19)$$

By substituting these equations into eqn (2.3) we obtain an expression for the *complex Fourier series*

$$
\begin{aligned}
x(t) &= 1/T \int_0^T x(p)\, dp + \sum_{k=1}^{n} \left[1/T \int_0^T x(p)\exp(-jk\omega_0)\, dp \right]\exp(jk\omega_0 t) \\
&\quad + \sum_{k=1}^{n} \left[1/R \int_0^T x(p)\exp(jK\omega_0 p)\, dp \right]\exp(-jk\omega_0).
\end{aligned}
\qquad (2.20)
$$

Equation (2.20) contains a succession of exponential terms extending from $k=1$ to $k=n$. The sign of k can be reversed in the second set of terms to obtain a simpler expression,

$$x(t) = \sum_{k=-t}^{k=n} 1/T \left[\int_0^T x(p)\exp(-jk\omega_0 p)\, dp \right]\exp(jk\omega_0 t)$$

or

$$x'(t) = \sum_{k=-n}^{k=n} A_k \exp(jk\omega_0 t). \qquad (2.21)$$

This represents a complex Fourier series for the signal, $x(t)$ limited to $2n - 1$ terms.

2.2 Fourier transformation

The Fourier series transformation, described earlier as a means of analysis in the frequency domain, has two major limitations that prevent its application to many time-series of practical interest. In the first case, it assumes that the time function is of *infinite duration,* whereas practical data are often transients with finite duration; and second, the assumption is made that the data are *periodic* over an unlimited extent, whereas practical data are usually non-periodic, as well as being limited in duration.

It is possible in many cases to represent non-periodic data by means of a *Fourier integral transform* (introduced in the previous chapter). This will be referred to as a *Fourier transform*—with it we may perform harmonic analysis and obtain power spectra, correlation functions, filtered data, coherence functions, and many other signal processing functions.

Intuitively we realize that it will be necessary to extend the range of integration to infinity in order to include the complete transient signal, since we can no longer assume that the data will repeat themselves outside the range of integration, previously limited to one fundamental period. Thus, we let k, $T \rightarrow \infty$ so that the fundamental frequency will tend towards zero, i.e. $f \rightarrow \delta f \rightarrow 0$. The Fourier coefficients will become continuous functions of frequency as the separation between the harmonics tends to zero and $k\omega_0 = (2\pi k)/T \rightarrow \omega$. Both k and $T \rightarrow \infty$ and k/T, now a continuous number, is the frequency f. We note that f and ω are related as $\omega = 2\pi f$.

Thus in this limiting condition eqn (2.18) becomes

$$\lim_{T \to \infty} A_k = \int_{-\infty}^{\infty} x(t)\exp(-j\omega_0 t)\,\mathrm{d}t. \qquad (2.22)$$

The right-hand side of eqn (2.22) is defined as the *complex Fourier transform* of $x(t)$, denoted by $X(f)$ as

$$X(f) = \int_{-\infty}^{\infty} x(t)\exp(-j\omega t)\,\mathrm{d}t. \qquad (2.23)$$

The absolute value of this function will yield the frequency amplitude content and the argument the phase content. Equation (2.21) now becomes

$$x(t) = \int_{-\infty}^{\infty} X(f)\exp(j\omega t)\,\mathrm{d}f \qquad (2.24)$$

which is known as the *inverse complex Fourier transform*, or *inverse transform* of the frequency series $X(f)$.

2.2.1 Some Fourier transforms

Certain characteristically shaped signals are met frequently in signal processing and other applications and it will be useful to review some of these briefly here.

(*a*) *A rectangular pulse of width T*

$$f_r(t) = 1 \quad \text{for} \quad -T/2 \leqslant t \leqslant T/2$$
$$= 0 \quad \text{otherwise} \tag{2.25}$$

(from eqn (2.23))

$$F_r(f) = \int_{-T/2}^{T/2} \exp(-\mathrm{j}2\pi ft)\,\mathrm{d}t$$
$$= [\exp(-\mathrm{j}\pi fT) - \exp(\mathrm{j}\pi fT)]/(-\mathrm{j}2\pi f)$$
$$= T(\sin(\pi fT))/(\pi fT). \tag{2.26}$$

$F_r(f)$ is plotted in Fig. 1.7(b) and is known as a *sinc function*. This is the rectangular window function referred to in Section 1.3.2.

(*b*) *An exponential function*

$$f_e(t) = \exp(\mathrm{j}2\pi ft) \tag{2.27}$$

From eqn (2.23) and taking into account the unlimited growth (or decay) of an exponential function

$$F_e(f) = \lim_{T \to \infty} \int_{-T/2}^{T/2} \exp(\mathrm{j}2\pi f_0 t)\exp(-\mathrm{j}2\pi ft)\,\mathrm{d}t$$
$$= \lim_{T \to \infty} [T(\sin(\pi(f-f_0)T)/(\pi(f-f_0)T)]. \tag{2.28}$$

The function inside the brackets is the sinc function of eqn (2.26).

As T goes to infinity so the peak shown in Fig. 1.7(b) increases and narrows, with first zero crossovers occurring at $f_0 \pm 1/T$. Hence at $T = \infty$, $F_e(f)$ becomes the *Dirac delta function*

$$F_e(f) = \delta(f - f_0) = \infty \quad \text{for} \quad f = f_0$$
$$= 0 \quad \text{otherwise.} \tag{2.29}$$

(*c*) *A shifted function of time*

$$f(\tau) = x(t - \tau). \tag{2.30}$$

Substituting a change of variable, $z = (t - \tau)$ in eqn (2.23)

$$F(\tau) = \int_{-\infty}^{\infty} x(t - \tau)\exp(-j2\pi ft)\,dt$$

$$= \int_{-\infty}^{\infty} x(z)\exp(-j2\pi f\tau)\exp(-j2\pi fz)\,dz.$$

Since $\exp(-j2\pi f\tau)$ does not vary with z this can be taken outside the transformation and we can write

$$F(\tau) = \exp(-j2\pi f\tau)X(f). \tag{2.31}$$

In other words if we take a signal $x(t)$ that is transformed to $X(f)$ and then consider a similar signal phase-shifted by τ to give $x(t - \tau)$, then the Fourier transform obtained is a linear modification of the amplitude of the modulus of the transform and no change of shape occurs. This is the *shifting theorem* for a sine/cosine function and has important consequences in arriving at the correlation of two signals through the Fourier transform, as we shall see later.

2.3 The discrete Fourier transform

In order to derive a form of Fourier transform for discrete data from the continuous form given in eqns (2.23) and (2.24) it is necessary to:
  (a) replace integrals by sums;
  (b) recognize that the limits of summation cannot be infinite.
We saw in Section 1.1 dealing with the discrete Fourier series that these limitations result in a discrete form of Fourier series representation that passes through all the sampled data values in a real discrete sequence of values x_i. We can see that in order to use these equations we must let the spectral series be complex, thus

$$X_n = (a_n + jb_n) \tag{2.32}$$

which would represent *both* positive and negative frequencies. The inverse discrete Fourier transform can then be written as

$$x_i = \sum_{n=-N/2}^{N/2} X_n \exp(j2\pi in/N) \tag{2.33}$$

($i = 1, 2, 3, \ldots, N$) where x_i and X_n are periodic functions. However, since the spectrum is given in complex form there are two spectral components generated for each real frequency. Consequently the summation of the pairs of components in the Fourier transform will result in a doubling of the amplitude of the spectral series produced. The Fourier transform must therefore include a scaling factor of $1/N$ as shown below

$$X_n = 1/N \sum_{i=-N/2}^{N/2} x_i \exp(-j2\pi in/N) \qquad (2.34)$$

Equations (2.33) and (2.34) form a Fourier transform pair suitable for expressing a discrete series. A simplified form of these equations can be obtained by noting that the series x_i and X_n are symmetrical for positive and negative values of N and include the zero value. We can therefore express the *discrete Fourier transform* (DFT) as

$$X_n = 1/N \sum_{i=0}^{N-1} x_i W^{in} \qquad (2.35)$$

$(i = 0, 1, \dots, N-1)$ and

$$x_i = \sum_{n=0}^{N-1} X_n W^{-in} \qquad (2.36)$$

$(n = 0, 1, \dots, N-1)$ for the *inverse discrete Fourier transform* (IDFT), where

$$W = \exp(-j2\pi/N) \qquad (2.37)$$

The convention, $x_i \leftrightarrow X_n$, is used to represent the transform pair with the meaning that x_i will transform to X_n via eqn (2.35) and similarly X_n will transform to x_i via eqn (2.36).

2.3.1 Some properties of the DFT

Succeeding chapters will consider the computer implementation and application of the DFT. Here a summary is given of its essential properties, most of which were discussed in earlier sections.

(*a*) *The DFT is a linear transform* If

$$x_i \leftrightarrow X_n \quad \text{and} \quad z_i = ax_i + by_i \quad \text{then} \quad Z_n = aX_n + bY_n. \qquad (2.38)$$

(*b*) *The DFT is periodic* If a continuous signal is sampled with a sampling frequency of f_s Hz, where $f_s = 1/h$ and h is the sampling period, then this sampled signal has a periodic spectrum that repeats at intervals of the sampling frequency f_s.

(*c*) *The DFT is complex conjugate* Given $x_i = x_0, x_1, \dots, x_{N-1}$ as a real-valued sequence such that $N/2$ is an integer and $x_i \leftrightarrow X_n$, then

$$X(N/2 + k) = X^*(N/2 - k) \qquad (2.39)$$

$(k = 0, 1, \dots, N/2)$.

(*d*) *The DFT is shift-invariate*   This was shown earlier for the continuous transform. Thus for $x_i \leftrightarrow X_n$ and $z_i = x_{i+\tau}$ then

$$Z_n = X_n W^{-n\tau} \qquad (2.40)$$

($\tau = 0, 1, 2, \ldots, N-1$).

(*e*) *The DFT and the correlation process are transform-related*   If $x_i \leftrightarrow X_n$, $y_i \leftrightarrow Y_n$ then their cross-correlation is, from eqn (1.20)

$$R_{xy}(\tau) = 1/(N\tau) \sum_{i=1}^{N-1} x_i y_{i+\tau} \qquad (2.41)$$

($\tau = 0, 1, \ldots, N-1$) and may be transformed to give

$$R_{xy} \leftrightarrow R_n = X_n^* Y_n \qquad (2.42)$$

($n = 0, 1, \ldots, N-1$). For the special case of auto-correlation when $x_i = y_i$ then eqn (2.42) reduces to

$$R_n' = |X_n|^2 \qquad (2.43)$$

which leads to the Wiener–Khintchine theory already noted in Section 1.5.2 as a route to the evaluation of the power spectral density.

2.4 Non-sinusoidal functions

A criterion for successful representation is, as stated earlier, that the constituent functions belong to an orthogonal set. Very many alternative function sets can be used. Some of these are non-periodic in form and of interest to the mathematician. Examples are the Legendre and Laguerre polynomials, which can be made orthogonal by multiplication by a weighting function. Others are capable of expansion directly in terms of a harmonic series and many of these are amenable to fast domain transformation procedures.

   In terms of accuracy in waveform representation the discrete Karhunan–Loève [3] series is optimal in the mean-square sense. However this is not an easy transform to calculate using a digital computer, and no effective fast transform algorithm exists. For this reason the sub-optimal rectangular functions of Walsh and Haar and their derivatives form the principal non-sinusoidal functions and are particularly applicable to digital computer implementation [4, 5, 6].

2.4.1 Walsh and Haar series

Walsh functions form an ordered set of rectangular waveforms taking only two amplitude values, +1 and −1, defined over a limited time interval. The first eight functions, arranged in order of increasing number

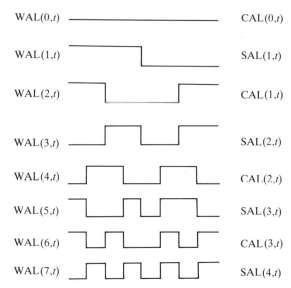

Fig. 2.2   A set of Walsh functions.

of crossovers for the zero axis, are shown in Fig. 2.2. Two arguments are required for complete definition; a time period t, and an ordering number n, related to frequency. The function may be expressed as

$$WAL(n, t). \qquad (2.44)$$

A further notation is used to classify the Walsh functions in terms of even and odd waveform symmetry

$$WAL(2k, t) = CAL(k, t) \qquad (2.45)$$
$$WAL(2k - 1, t) = SAL(k, t)$$

$(k = 1, 2, \ldots, (N/2 - 1)$. This defines two further Walsh series having close similarities with the cosine and since series, as we shall see later.

The ordering number n gives the periodic repetition rate of the waveform and is termed the *sequency* for the function. It is defined as 'one half of the average number of zero crossings per unit time interval' [4]. From this we see that frequency can be regarded as a special measure of sequency applied to sinusoidal waveforms only. In eqn (2.45) k is seen to represent sequency in the CAL or SAL ordering.

The discrete Walsh function series corresponding to a Fourier series in Fourier analysis is expressed as

$$f(t) = a_0 WAL(0, t) + \sum_{i=1}^{N/2-1} \sum_{j=1}^{N/2-1} [a_i SAL(i, t) + b_j CAL(j, t)]. \quad (2.46)$$

Here a_i and b_j are coefficients expressing the series in the sequency domain. The derivation of these coefficient series is referred to as decomposition into the spectral components of $f(t)$ although these are now no longer sinusoidal in form. It is important to note that although it is possible to combine sine and cosine elements into a single complex variable, expressing the same frequency, this is not possible with Walsh functions, due to the absence of a similar shift theorem to that found with the Fourier function (see Section 2.2.1).

The Walsh series is orthogonal and it can be shown that for two Walsh functions, WAL(m, t) and WAL(n, t),

$$\sum_{t=0}^{N-1} \text{WAL}(m, t)\text{WAL}(n, t) = N \quad \text{for } n = m$$

$$= 0 \quad \text{for } n \neq m. \quad (2.47)$$

Thus the Walsh functions form an orthogonal set which can be normalized through division by N to form an orthonormal system.

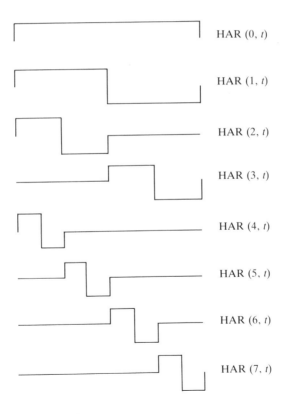

Fig. 2.3 A set of Haar functions.

Using the Walsh function series any complex waveform can be represented by superposition of a number of elements of the series, having appropriate amplitudes, in an exactly analogous manner to the familiar Fourier synthesis using sine/cosine functions. It will be found that fewer Walsh functions are required to represent a discontinuous waveform, compared with synthesis using sine/cosine functions, and that the converse holds true for smoothly varying waveforms [7]. This brings out an important difference between the two sets of functions, indicating their respective roles in processing and control applications.

The Haar function series, shown in Fig. 2.3, is also formed from a set of periodic square waves. The amplitude values of these square waves do not have uniform value, as with Walsh waveforms, but assume a limited set of values, $0, \pm 1, \pm\sqrt{2}, \pm 2, \pm 2\sqrt{2}, \pm 4$, etc. They may be expressed in a similar manner to the Walsh functions as

$$\text{HAR}(n, t). \tag{2.48}$$

If we consider the time base to be defined as $0 < t < 1$, then following the simplified definition suggested by Kremer [8], we can write

$$
\begin{aligned}
\text{HAR}(0, T) &= 1 \quad \text{for } 0 \leqslant T \leqslant 1 \\
\text{HAR}(1, t) &= 1 \quad \text{for } 0 \leqslant t < 1/2 \\
&= -1 \quad \text{for } 1/2 \leqslant t < 1/2 \\
\text{HAR}(2, t) &= \sqrt{2} \quad \text{for } 0 \leqslant t < 1/4 \\
&= -\sqrt{2} \quad \text{for } 1/4 < t \leqslant 1/2 \\
&= 0 \quad \text{for } 1/2 < t \leqslant 1 \\
\text{HAR}(3, t) &= 0 \quad \text{for } 0 < t \leqslant 1/2 \\
&= \sqrt{2} \quad \text{for } 1/2 \leqslant t < 3/4 \\
&= -\sqrt{2} \quad \text{for } 3/4 < t \leqslant 1 \\
&\cdots \\
&\cdots \\
\text{HAR}(2^p + nt) &= \sqrt{2^p} \quad \text{for } n/2^p \leqslant t < (n + 1/2)/2^p \\
&= -\sqrt{2^p} \quad \text{for } (n + 1/2)/2^p \leqslant t < (n + 1)2^p \\
&= 0 \quad \text{for elsewhere}
\end{aligned}
\tag{2.49}
$$

$(p = 1, 2, \ldots, \log_2 N \quad n = 0, 1, \ldots, 2^p - 1 \quad N = 2^p)$. This allows a sequential numbering system analogous to that adopted by Walsh for his function series.

We see from Fig. 2.3 that the first two of the Haar functions are identical to WAL$(0, t)$ and WAL$(1, t)$. The next function, HAR$(2, t)$ is simply HAR$(1, t)$ shifted into the left-hand half of the time-base and modified in amplitude to $+2$. The next function, HAR$(3, t)$ is identical but shifted into the right-hand half of the time base. In general all members of the same function subset are obtained by a lateral shift of the first member along the time axis by an amount proportional to its length. The essential characteristic of the Haar function is seen as a constant value everywhere except in one sub-interval, where a double step occurs.

From the definition given above it can be seen that the Haar functions are also orthogonal, i.e.

$$\sum_{t=0}^{N-1} \text{HAR}(m, t)\text{HAR}(n, t) = N \quad \text{for } n = m$$

$$= 0 \quad \text{for } n \neq m. \tag{2.50}$$

There is no theorem analogous to the shift or addition theorems (Sections 2.2.1 and 2.4.3) found with Fourier and Walsh series, respectively.

2.4.2   Transformation of Walsh and Haar series

A finite discrete Walsh transform pair can be stated for a time series x_i comprised of N samples. Thus we have

$$X_n = 1/N \sum_{i=0}^{N-1} x_i \text{WAL}(n, i) \quad n = 0, 1, 2, \ldots, N-1 \tag{2.51}$$

$$x_i = \sum_{n=0}^{N-1} X_n \text{WAL}(n, i) \quad i = 0, 1, 2, \ldots, N-1. \tag{2.52}$$

The transform and its inverse may be obtained by matrix multiplication using a digital computer. Since the matrices are symmetrical for the Walsh transform (unlike the Fourier transform) then both transform and inverse transform are identical, except for a scaling factor $1/N$. Comparing eqn (2.51) with the corresponding discrete Fourier transform,

$$X_f + 1/N \sum_{i=0}^{N-1} x_i \exp(-\text{j}2\pi i f/N) \tag{2.53}$$

$(f = 0, 1, 2, \ldots, N-1)$. We see that while $\text{WAL}(n, i)$ is real and limited to values ± 1, the exponential kernel in eqn (2.53) is complex and can assume N different values. As a direct consequence of this the Walsh transform proves considerably easier and faster to calculate using digital methods.

The discrete Haar transform (DHT) and its inverse are given by

$$X_n = 1/N \sum_{i=0}^{N-1} x_i \, \text{HAR}(n, i/N) \tag{2.54}$$

$$x_i = \sum_{n=0}^{N-1} X_n \, \text{HAR}(n, i/N) \tag{2.55}$$

$(i, n = 0, 1, 2, \ldots, N - 1)$. Unlike the Walsh transform the matrix is not symmetric, so separate transform operations are required for transformation and inverse transformation.

As will be seen later there are considerable differences in calculation time using the Walsh or Haar fast discrete transformations. While the Haar transform is faster to compute, its characteristics are not so useful as the Walsh transform for general spectral domain work, although the shape of the Haar waveform does provide better matching in correlation applications involving abrupt level transitions such as, for example, edge detection in image processing [9].

2.4.3   *Problems and special characteristics*

The discrete Fourier transform is invariant to the phase of the input signal so that the same spectral decomposition can be obtained independently of the phase or circular time shift of the input signal. This is not the case with the Walsh transform. The effect is one of variation in shape and location of the dominant spectral characteristics of the signal. It is of less importance where the sum of the squares of pairs of transformed coefficients of the same sequence are taken, as with power spectrum derivation, but will account for minor variations between spectra unless the signals are obtained at the same time or adjusted for zero phase shift [10].

The reason for this variation in composition of the Walsh spectrum is found in the addition relationship for the Walsh functions, namely,

$$\mathrm{WAL}(k, t)\mathrm{WAL}(p, t) = \mathrm{WAL}(k \oplus p, t) \qquad (2.56)$$

where $\oplus$ represents modulo-2 addition.

This corresponds to the set of trigonometric relationships

$$2 \cos kt \cos pt = \cos(k - p)t + \cos(k + p)t, \ldots, \text{etc.}$$

with the important difference that a *shift theorem* for Walsh functions *does not exist*, so that while the product of two Fourier transforms can be transformed to obtain a convolution of the two original time series, a similar result is *not* obtained with the Walsh transform. The absence of a Walsh shift theorem prevents fast correlation via the Walsh transform and effective digital filtering other than through the Wiener filtering described in Section 1.3.3.

The Walsh transform does however provide an equivalent Periodogram method of power spectral analysis suitable for transient or discontinuous waveforms. To carry this out the sum of the squares of the CAL and SAL function coefficients of the same sequency are added for each spectral coefficient in a manner similar to that used in Fourier analysis. Reference to Fig. 2.2 and eqn (1.29) will show that $N/2 + 1$ spectral points will be obtained with the first and the last consisting of the square of the CAL or SAL function only, i.e.

$$P_{(0)} = X_c^2(0, t)$$

$$P_{(k)} = X_c^2(k, t) + X_s^2(k, t)$$

$$P_{(N/2)} = X_s^2(N/2, t) \tag{2.57}$$

$(k = 1, 2, \ldots, (N/2 - 1))$, where $X_c(k, t)$ is the Walsh transform involving CAL(k, t) and $X_s(k, t)$ is the Walsh transform involving SAL(k, t).

A similar Periodogram definition for the Haar power spectrum is difficult to use with the definition given by eqn (2.49), since the equivalent periodicity of a number of adjacent functions can be identical. If we define the effective sequency of the Haar function series as 'one half the average number of zero crossings per unit time interval' as we did for the Walsh series, then it will be seen that the Haar functions fall into discrete groups, each member of a group having the same effective sequency as other members of the same group (see Table 2.1). Using this definition a power spectrum may be defined, which can be considered as analogous to the sequency or frequency spectrum by taking the normalized value of the squares of the line spectra that fall within each grouping. The spacing of sequency values obtained will not be uniform and only a small number of spectral values will be obtained. The sparsity of calculated values may however be considered acceptable in some circumstances [8].

Table 2.1 Haar spectral groups

	Sequency group
HAR$(0, t)$	1
HAR$(1, t)$	2
HAR$(2, t)$, HAR$(3, t)$	3
HAR$(4, t)$, HAR$(5, t)$ HAR$(6, t)$, HAR$(7, t)$	4
HAR$(8, t)$, HAR$(9, t)$ HAR$(10, t)$, HAR$(11, t)$ HAR$(12, t)$, HAR$(13, t)$ HAR$(14, t)$, HAR$(15, t)$	5

A summary of the mathematical characteristics of the Walsh and Haar series is given in Table 2.2, where they are compared with the Fourier series.

2.5 Two-dimensional transformation

Discrete orthogonal transformations in two dimensions are required for a number of applications, the most obvious one being that of image

Table 2.2 Summary of mathematical characteristics for orthogonal series

	Fourier	Walsh	Haar
Function	$\text{SIN}(\omega, t)$ $\text{COS}(\omega, t)$	$\text{WAL}(n, t)$ $\text{CAL}(k, t), \text{SAL}(k, t)$	$\text{HAR}(n, t)$
Discrete Fourier series	$x_i = a_0 + \sum_{k=1}^{N/2} a_k \cos(k\omega_0 t)$ $\quad + \sum_{k=1}^{N/2} b_k \sin(k\omega_0 t)$	$x_i = a_0 \text{WAL}(0, 0)$ $\quad + \sum_{m=1}^{N/2} a_m \text{SAL}(m, i)$ $\quad + \sum_{p=1}^{N/2} b_p \text{CAL}(p, i)$	$x_i = \text{HAR}(0, 0) + \sum_{n=1}^{n-1} \text{HAR}(n, i)$
Discrete transform	$X_n = 1/N \sum_{i=0}^{N-1} x_i W^{in}$ $\quad (i = 0, 1, \ldots, N-1)$	$X_n = 1/N \sum_{i=0}^{N-1} x_i \text{WAL}(n, i)$	$X_n = 1/N \sum_{i=0}^{N-1} x_i \text{HAR}(n, i/N)$
Discrete inverse transform	$x_i = \sum_{n=0}^{N-1} X_n W^{-in}$ $\quad (n = 0, 1, \ldots, N-1)$ $\quad W = \exp(-j2\pi/N)$	$x_i = \sum_{n=0}^{N-1} X_n \text{WAL}(n, i)$	$x_i = \sum_{n=0}^{N-1} X_n \text{HAR}(n, i/N)$ $\quad (i, n = 0, 1, 2, \ldots, N-1)$
Orthogonal property	$\sum_{i=0}^{N-1} \exp(j2\pi ni/N)\exp(j2\pi mi/N)$ $\quad = 1 \;\; \text{for } n = m$ $\quad = 0 \;\; \text{for } n \neq m$	$\sum_{t=0}^{N-1} \text{WAL}(m, t)\text{WAL}(n, t)$ $\quad = 1 \;\; \text{for } n = m$ $\quad = 0 \;\; \text{for } n \neq m$	$\sum_{t=0}^{N-1} \text{HAR}(m, t)\text{HAR}(n, t)$ $\quad = N \;\; \text{for } n = m$ $\quad = 0 \;\; \text{for } n \neq m$

processing [11]. A digitized discrete spatial image may be expressed as a series of sub-images or pixels $x_{i,k}$, where $i = 0, 1, \ldots, N-1$ and $k = 0, 1, \ldots, M-1$. A two-dimensional DFT evaluated at discrete angular frequency values

$$\omega_n = n(2\pi/N) \quad \text{and} \quad \omega_m = m(2\pi/N) \qquad (2.58)$$

is given as

$$X_{n,m} = 1/MN \sum_{i=0}^{N-1} \sum_{k=0}^{M-1} x_{i,k} \exp(-j(\omega_n i + \omega_m k)) \qquad (2.59)$$

and the inverse transform given by

$$x_{i,k} = \sum_{n=0}^{N-1} \sum_{m=0}^{M-1} X_{n,m} \exp(j(\omega_n i + \omega_m k)). \qquad (2.60)$$

Even though $x_{i,k}$ is a real positive function, its transform $X_{n,m}$ is generally complex; thus while the image contains NM components, the transform contains $2NM$ components, the real and imaginary or magnitude and phase components of each spatial frequency. However, since $x_{i,k}$ is a real positive function, $X_{n,m}$ exhibits a property of conjugate symmetry, i.e.

$$X_{n,m} = -X_{-n,-m}. \qquad (2.61)$$

Consequently

$$X_{n,m} = X^*_{-n,-m}. \qquad (2.62)$$

Due to the conjugate symmetry property of the Fourier transform it is only necessary to consider the samples of one-half of the transform plane; the other half can be reconstructed from the half-plane samples, thus enabling the Fourier transform of an image to be adequately described by NM data components.

As in the one-dimensional case we find that for a two-dimensional representation to be valid the field must be periodic. We must therefore consider the regional image periodically repeated in the horizontal and vertical directions, and this needs to be considered in subsequent processing of the image.

In a practical calculation of the two-dimensional DFT it is not necessary to evaluate directly the double summation form of the transform equation as shown in eqn (2.59). If instead the two-dimensional data matrix $x_{i,k}$ is subject to two separate sets of DFT operations, one set dealing with the rows of the data matrix and a second set carrying out a transformation on the resulting columns of coefficient values, this reduces the problem to a multiple single-dimensional transform operation. The process will be considered later (Section 5.6) following description of fast transformation algorithms.

While the two-dimensional Fourier transform possesses many useful analytical properties, it has two major drawbacks; complex rather than real number computations are necessary and the rate of convergence is low. This latter disadvantage is significant in image work, where very large databases need to be handled. Other transforms that do not have these major drawbacks are the cosine and sine transforms and the non-sinusoidal Walsh and Haar transforms.

2.5.1   The cosine transform

The Fourier series representation of a continuous real and symmetric function contains only real coefficients corresponding to the cosine terms of the series. This condition can be realized with an image field and a compact discrete cosine transform obtained that is capable of fast implementation [12].

The *discrete cosine transform* (DCT) is defined as

$$X_n = 1/N \sum_{i=0}^{N-1} x_i \cos[(n\pi(2i+1))/2N] \qquad (2.63)$$

and its inverse as

$$x_i = 2 \sum_{i=0}^{N-1} X_n \cos[(n\pi(2i+1))/2N] - X_0 \qquad (2.64)$$

where X_0 is the average value and $i, n = 0, 1, 2, \ldots, N-1$.

The cosine transform can be derived directly from the complex Fourier transform as

$$X_n = \mathrm{Re}\left[\exp(-(jn\pi)/2N) \sum_{i=0}^{2N-1} x_i W^{ni} \right] \qquad (2.65)$$

where $W^{ni} = \exp(-j2\pi)/N$ and Re[.] denote the real part of the term enclosed.

From this relationship it is easy to see that the DFT forms a route to the DCT. However, as we shall see later, other derivations can be obtained that are more economical in computer time.

The value of the cosine transform for such applications as image processing and data compression lies in its good variance distribution and low rate distortion function. This results in efficient energy compaction, where the transform coefficients containing the largest variance are found to be contained in about a quarter of the transformed image. This is virtually the same energy compaction found with the Karhunen–Loève transform, known to be optimal in the mean-square sense. This may be seen from Fig. 2.4 (with acknowledgement to the IEEE and Dr. W. K. Pratt) where the MSE performances of various transforms are compared

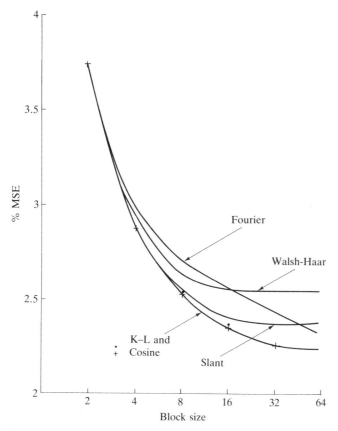

Fig. 2.4 Comparative performance of orthogonal transformations. (After Pratt *et al.* [13].)

[13]. In the following chapters the derivation of fast transformations for the cosine and other orthogonal transformations will be considered in further detail.

2.5.2 Walsh and Haar transforms

For symmetric sets of sampled values $x_{i,k}$ ($N = 2^p$) the two-dimensional Walsh and Haar transforms may be carried out as with the two-dimensional Fourier transform, i.e. through a single-dimensional transform process applied sequentially to the rows and columns of the matrix array. The two-dimensional array, $x_{i,k}$ of N^2 values is given by,

$$X_{n,m} = 1/N^2 \sum_{i=0}^{N-1} \sum_{k=0}^{N-1} x_{i,k} \text{WAL}(ni) \text{WAL}(mk) \qquad (2.66)$$

and the inverse transformation by

$$x_{i,k} = \sum_{m=0}^{N-1} \sum_{n=0}^{N-1} X_{n,m} \text{WAL}(n, i) \text{WAL}(m, k) \qquad (2.67)$$

using the symmetry of the discrete functions.

In a similar manner the two-dimensional Haar transform pair can be stated as,

$$X_{m,n} = 1/N^2 \sum_{i=0}^{N-1} \sum_{j=0}^{N-1} x_{i,j} \text{HAR}(n, i/N) \text{HAR}(m, j/N). \qquad (2.68)$$

Other approaches to the two-dimensional Haar transform are given by Shore [14]; these are more applicable to image work. Shore uses optical techniques in which transformation is via a transparent coded mask. Further analog methods of transformation by optical means have been applied directly to both Walsh and Haar two-dimensional representation. These are generally also mask-coded techniques and not too difficult to arrange for the non-sinusoidal functions, since only a small number of transmission levels are required in the mask representation. Several examples are given in Beauchamp [5].

References

1. Wiener, N. *The extrapolation, interpolation and smoothing of stationary time series.* MIT Press, Cambridge, MA. (1949).
2. Champeney, D. C. *Fourier transforms and their physical applications* pp. 221–2. Academic Press, London (1973).
3. Davenport, W. B. and Root, W. L. *An introduction to the theory of random signals and noise.* McGraw-Hill, New York (1958).
4. Harmuth, H. F. *Transmission of information by orthogonal functions.* Springer, Heidelberg (1972).
5. Beauchamp, K. G. *Applications of Walsh and related functions.* Academic Press, London (1984).
6. Maqusi, M. *Applied Walsh analysis.* Heyden Press, London (1981).
7. Beauchamp, K. G. Waveform synthesis using Fourier and Walsh series. *Proceedings Theory and applications of Walsh functions.* Hatfield Polytechnic, Hatfield, UK (1973).
8. Kremer, H. Algorithms for the Haar function and the fast Haar transform. *Proceedings Theory and applications of Walsh functions.* Hatfield Polytechnic, Hatfield, UK (1971).
9. Rosenfield, A. and Thurston, M. Edge and curve detection for visual scene analysis. *IEEE Trans. Comput.* **C-20,** 562–9 (1971).
10. Beauchamp, K. G. The Walsh power spectrum. *Proc. NEC,* (Chicago) **27,** 377–82 (1972).
11. Andrews, H. C. *Computer techniques in image processing.* Academic Press, New York (1970).

12. Ahmed, N., Natarajan, T. and Rao, K. R. Discrete cosine transform. *IEEE Trans. Comput.* **C-23,** 90–3 (1974).
13. Pratt, W. K. Chen, W. H., and Welch, L. R. Slant transform image coding. *IEEE Trans. Commun.* **COM-22** (8), 1075–93. (1974).
14. Shore, J. E. A two-dimensional Haar-like transform, NDRL Report No. 71472, AD755433 (1973).

3

MATRICES AND OTHER ESSENTIAL MATHEMATICS

3.1 Introduction

The mathematics required to understand and apply the principles of domain transformation is not difficult. It is however complicated by the great variety of different techniques and interpretations found in the published application papers. This chapter has been written with the aim of presenting a concise interpretation of these principles and of setting out a résumé of the essential mathematical ideas.

The two basic notions dominating the subject are those of linear mathematics and matrix algebra [1–4]. It is in fact profitable in many ways to consider a digital transform as an operation in linear algebra, particularly when examining symmetries in the transform with the aim of reducing the amount of computation required.

In this book a matrix formulation of the transformation techniques is generally given and associated with the use of signal flow diagrams as a statement of algorithm design. In addition to these purely mathematical topics some familiarity with the ideas of discrete data manipulation is given, examples being modulo-2 arithmetic, bit reversal, and code conversion.

A limited and non-rigorous treatment of these topics is given below, commencing with an introduction to linear mathematics.

3.2 Linear mathematics

3.2.1 Scalars, vectors, and functions

Scalars represent simply the measurement of one entity in terms of another, where the relationship is a linear one. Thus,

$$y = Ax + B \qquad (3.1)$$

is a linear relationship giving rise to *scalar* values for y in terms of x, where A and B have fixed (constant) values. A and B may change from example to example but they do not change with x.

Of more relevance to our studies are sets of scalar quantities having different values and arranged as a group of N numbers. These are known as *vectors* which may be arranged as a *column vector*

$$X_c = \begin{bmatrix} x_1 \\ x_2 \\ \vdots \\ x_N \end{bmatrix}$$

(3.2)

or in horizontal form as a *row vector*

$$X_r = [x_1, x_2, \ldots, x_N].$$

(3.3)

Here the quantities x_i (where $i = 1, 2, \ldots, N$) are called the *components* of x while N is called the *dimension* of vector x.

A set of rules for manipulating vector quantities must obey the following notions of linearity.

1. Two vectors x and y are said to be equal if their constituent components are also equal, i.e. $x_i = y_i$ for $i = 1, 2, \ldots, N$.

2. Vector addition can be expressed as

$$X + Y = \begin{bmatrix} x_1 + y_1 \\ x_2 + y_2 \\ \vdots \\ x_N + y_N \end{bmatrix}.$$

(3.4)

3. The rules of commutativity, $x + y = y + x$, and associativity, $x + (y + z) = (x + y) + z$, also apply.

4. Multiplication of a vector X by a scalar A is defined as

$$AX = XA = \begin{bmatrix} Ax_1 \\ Ax_2 \\ \vdots \\ Ax_N \end{bmatrix}.$$

(3.5)

A similar set of rules apply to *functions* of x or y, i.e. $f(x)$ and $f(y)$, which are often expressed as functions of time $x(t)$ and $y(t)$. A function can be (and usually is) a continuous scalar relationship between a continuously variable value of one quantity against another. A vector, on the other hand, is a discrete set of scalar quantities which can nevertheless have a functional relationship between the two quantities, (e.g. a sampled set of function values).

A most important scalar function of two vectors x and y is known as the *inner product* and is met with in one form or another throughout domain transformation and other signal processing operations. The inner product of two vectors X and Y is

$$XY = \sum_{i=1}^{N} x_i y_i$$

(3.6)

and that between two real functions is

$$\int x(t)y(t)\,\mathrm{d}t. \tag{3.7}$$

If $X = Y$ then the inner product is

$$XX = \sum_{i=1}^{N} |x_i|^2. \tag{3.8}$$

The square root of XX is called the *norm* of X, denoted as $|X|$. Clearly $|X|$ is zero only when x_i is zero for every value of i. Consequently the norm indicates how significant a quantity a vector is.

Many of the vectors we will be considering are complex quantities, i.e.

$$x = A + \mathrm{j}B \tag{3,9}$$

where A and B are any positive or negative real numbers and $\mathrm{j} = \sqrt{-1}$.

The *complex conjugate, x^** of any complex number x is given as

$$x^* = A - \mathrm{j}B. \tag{3.10}$$

If x is a real number then $x^* = x$. (We used this same symbol earlier to represent convolution but there should be no difficulty identifying the two definitions in a given context).

Where x_i and y_i are complex number values contained within vectors X and Y, the inner product is defined as

$$XY = x_1^* y_1 + x_2^* y_2 + , \ldots , + x_N^* y_N$$
$$= \sum_{i=1}^{N} x_i^* y_i \tag{3.11}$$

and we can write

$$X \cdot Y = X^* \cdot Y^*$$
$$X(Y + Z) = X \cdot Y + X \cdot Z$$
$$(Y + Z)X = Y \cdot X + Z \cdot X. \tag{3.12}$$

A concept useful in considering matrix transform manipulation given later (in Chapter 4) is that of a *modulus*. Consider the linear expression $a = br + s$ where all values are integers. We can regard s as a remainder obtained when a is divided by b r times. Here b is fixed and called a modulus. For a given value of r then we could have an infinite number of values of a for a given r all of which result in the same remainder s. All these choices for a are said to be *congruent* with respect to modulus b, and s is called the *residue modulo-b*. For example, if $b = 5$ and a has values of 14, 9, or 24 then each gives 4 as a remainder r and all of these values of a are congruent modulo-6.

3.2.2 Orthogonality

We are now in a position to look at the property of orthogonality between sets of vectors. Two row (or column) vectors X, Y are orthogonal if the inner vector product is zero, i.e.

$$X \cdot Y = Y \cdot X = 0$$

that is

$$\sum_{i=1}^{N} x_i^* y_i = \sum_{i=1}^{N} y_i^* x_i = 0 \tag{3.13}$$

or in the case of real positive numbers for X, Y

$$\sum_{i=1}^{N} x_i y_i = 0 \tag{3.14}$$

$(i = 1, 2, \ldots, N)$. When the vectors are equal, however, (i.e. $x_i = y_i$) then this orthogonality vanishes and is replaced by a linear relationship

$$\sum_{i=1}^{N} x_i x_i = K \tag{3.15}$$

and for most purposes K is arranged to be 1, when the vector series is said to be *orthonormal*.

The orthogonal relationships were summarized for an orthonormal series of continuous functions in Section 2.1 through eqn (2.1).

This property of orthogonality permits the accurate representation of a complicated function by superposition of a set of simple functions and allows the identification of one particular vector or function from a combined set of vectors, i.e. signal synthesis and analysis [5].

3.2.3 Linear operators

Equation (3.1) expresses y in terms of a single value of x. Frequently the situation arises where y contains multiple values of x and each value of y is linearly related to each value of x. Different values of y may also have different coefficients. Thus we can have

$$y_k = \sum_{i=1}^{N} A_{ik} x_i. \tag{3.16}$$

Similarly we have for a function of t

$$y(t) = \int A(t, t') x(t') \, dt'. \tag{3.17}$$

The set of coefficients in eqn (3.16) form a *matrix* which is simply a two-dimensional table of numbers. The set of coefficients in eqn (3.17) form a *kernel*. Both matrices and kernels are known as *linear operators* because they convert a vector or a function x into another of y by means of linear arithmetic. Equations (3.16) and (3.17) can be expressed simply as $y = A \cdot x$, where A is the linear operator (matrix or kernel) and x is the operand.

An important operator, known as the *identity operator* and denoted as I, has the property of leaving the operand unchanged. For eqn (3.16) the identity operator is the Dirac impulse function that we met earlier in Section 1.1.1 and we can write $y = Ix = x$. The table of numbers formed by A_{ik} is called the *unit matrix*. It has 1s on the matrix diagonal and all elements off the diagonal are 0. Since the introduction of an identity operator in a matrix expression does not change the vector values for y its use can enable a matrix factorization to be achieved, which we see later is a valuable technique for fast transform evaluation [6].

For certain operators A we can find an operator B such that applying A and B in succession leaves the operand unchanged, i.e.

$$B \cdot A = I. \tag{3.18}$$

B is called the *inverse* of A.

Where an inverse exists then the operator is called a *non-singular* operator. An example here is the Walsh matrix operator, permitting the development of a transform that can be applied twice in succession to recover the original time domain vector. We say that such transforms 'contain their own inverse transform'.

It is not possible to find an inverse for every operator. Operators for which an inverse cannot be found are called *singular* operators. Using these destroys part of the information, thus making reversion back to the original data vector impossible. Note that even if a non-singular operator is found it may only be possible to apply this in one order of precedence, i.e. AB is not necessarily equal to BA.

Two further features of matrix operators are worth noting:

1. An operator is *symmetric* if

$$A_{ik} = A_{ki} \quad \text{or} \quad A(n, m) = A(m, n). \tag{3.19}$$

2. A is known as a *Hermitian operator* if

$$A_{ik} = A_{ki}^* \quad \text{or} \quad A(n, m) = [A(m, n)]^*. \tag{3.20}$$

3.3 Matrix algebra

The matrix is defined by eqn (3.16) and considered here as a linear operator. Matrix algebra provides a useful method for systematizing both the theoretical and practical aspects of several computing procedures and is the main tool for transform algorithm derivation [6, 7].

Row vectors and column vectors can be considered as single-dimensional matrices and their scalar product will give a two-dimensional matrix. However the result is only meaningful if the row and column have the same number of terms, i.e.

$$XY = [x_1, x_2 \ldots x_n] \begin{bmatrix} y_1 \\ y_2 \\ \vdots \\ y_n \end{bmatrix}$$

$$= x_1 y_1 + x_2 y_2 \ldots x_n y_n$$

$$= [x_i y_i] = A. \tag{3.21}$$

Similarly two matrices can be multiplied together only if the number of columns in the first is equal to the number of rows in the second. Then the element in the ith row and the jth column is the scalar product of the ith row of the first matrix with the jth column of the second matrix. Thus the product of two matrices A and B can be written

$$[a_{ij}][b_{jk}] = [c_{ik}] \tag{3.22}$$

and

$$c_{ik} = \sum_{j=1}^{n} a_{ij} b_{jk}. \tag{3.23}$$

For example

$$\begin{bmatrix} a_{11} & a_{12} \\ a_{21} & a_{22} \\ a_{31} & a_{32} \end{bmatrix} \begin{bmatrix} b_{11} & b_{12} \\ b_{21} & b_{22} \end{bmatrix} = \begin{bmatrix} a_{11}b_{11} + a_{12}b_{21}, \ a_{11}b_{12} + a_{12}b_{22} \\ a_{21}b_{11} + a_{22}b_{21}, \ a_{21}b_{12} + a_{22}b_{22} \\ a_{31}b_{11} + a_{32}b_{21}, \ a_{31}b_{12} + a_{32}b_{22} \end{bmatrix}. \tag{3.24}$$

The order in which matrices are multiplied together is thus quite important. Whilst the products AB and BA may both exist they will in general be quite different and no very useful connection between the two will exist. Thus the commutative law does not hold for matrix multiplication $(AB \neq BA)$. The distributive and associative laws are valid however, e.g.

$$(A + B)C = AC + BC$$

$$A(BC) = (AB)C = ABC. \tag{3.25}$$

A significant class of matrices are the *diagonal* matrices, $D = [d_{ij}]$. These consist of a square matrix (same number of rows and columns) in which the d_{ij} are all zero, except $d_{11}, d_{22}, \ldots, d_{nn}$, which are elements along the principal diagonal. Thus

$$A = \begin{bmatrix} a & 0 & 0 \\ 0 & b & 0 \\ 0 & 0 & c \end{bmatrix} \tag{3.26}$$

is a *right-diagonal matrix* and

$$B = \begin{bmatrix} 0 & 0 & a \\ 0 & b & 0 \\ c & 0 & 0 \end{bmatrix} \tag{3.27}$$

is a *left-diagonal matrix*.

The zeros are often omitted in the representation. Thus for the *unit matrix* I, which we met earlier and which has only unit value elements, we can write

$$I = \begin{bmatrix} 1 & & \\ & 1 & \\ & & 1 \end{bmatrix}. \tag{3.28}$$

Clearly

$$AI = IA = A. \tag{3.29}$$

In matrix algebra I plays the same role as unity in ordinary algebra.

Two rearrangements of a matrix that are useful in matrix manipulations are the *transpose* of a matrix and the *inverse* of a matrix.

The transpose A^{T} of a matrix, A is defined by the property that if A is an $M \times N$ matrix whose (i, j)th element is $[a_{ij}]$, then A^{T} is an $N \times M$ matrix whose (i, j)th element is $[a_{ji}]$, i.e.

$$[a_{ij}]^{\mathrm{T}} = [a_{ji}]. \tag{3.30}$$

A *symmetrical* matrix is also a symmetrical operator, where

$$[a_{ij}] = [a_{ji}], \quad \text{i.e. } A^{\mathrm{T}} = A. \tag{3.31}$$

It is sometimes convenient to express a column vector in horizontal form and this can also be achieved by using the transpose concept. Thus for the column vector of eqn (3.2) we can write

$$x_c = [x_1, x_2, \ldots, x_N]^{\mathrm{T}}. \tag{3.32}$$

The inverse of a matrix may be stated in terms of a unit matrix. Given an $N \times N$ matrix, A there exists a matrix, B such that

$$AB = I$$

where, unlike the general rule of matrix multiplication,

$$BA = I.$$

This matrix, B is called the inverse of A and denoted as A^{-1}, so that we can write

$$A^{-1}A = AA^{-1} = I. \tag{3.33}$$

Earlier we met the terms singular and non-singular applied to linear operators. Since matrices are a particular form of linear operator the concept also applies here. Thus we find that not all matrices have an inverse form and the rule is that for the matrix to be non-singular the determinant for A must not be zero, i.e. if $\text{Det}\,A \neq 0$ then the matrix is *non-singular* and has an inverse. If $\text{Det}\,A = 0$ the matrix is called *singular* and an inverse does not exist.

3.3.1 Determinants

With any square matrix there is associated a number $\text{Det}(.)$ called the *determinant*, which is calculated from products of the elements of the matrix. Thus for a 2×2 matrix,

$$A = \begin{bmatrix} a & b \\ c & d \end{bmatrix} \quad \text{then Det}(A) = ad - bc. \tag{3.34}$$

Similarly for a 3×3 matrix,

$$B = \begin{bmatrix} a & b & c \\ d & e & f \\ g & h & i \end{bmatrix} \tag{3.35}$$

then the determinant is

$$\text{Det}(B) = A\begin{pmatrix} e & f \\ h & i \end{pmatrix} + B\begin{pmatrix} f & d \\ i & g \end{pmatrix} + C\begin{pmatrix} d & e \\ g & h \end{pmatrix}$$
$$= a(ei - hf) + b(fg - id) + c(dh - ge). \tag{3.36}$$

Determinants for larger square matrices are calculated in a similar manner.

The principal use for determinants is in solving simultaneous equations, but they will also be found used as an indicator of the characteristics of a matrix, such as that of singularity mentioned previously.

3.3.2 Unitary matrices

Many matrices describing transform algorithms are known as *unitary matrices*. A unitary matrix is a square matrix A having an inverse that is the complex conjugate of its transposed value, i.e.

$$A^{-1} = (A^*)^{\mathsf{T}} = (A^{\mathsf{T}})^*. \tag{3.37}$$

A real unitary matrix is called an *orthogonal matrix*. For such a matrix

$$A^{-1} = A^{\mathsf{T}}. \tag{3.38}$$

Finally if A is a symmetrical orthogonal matrix then

$$A^{-1} = A. \tag{3.39}$$

We are now in a position to state a series of important rules for unitary matrices:

1. The determinant of a unitary matrix has the absolute value of 1.
2. The column and row vectors of a unitary matrix are mutually orthogonal unit vectors.
3. The inverse and transpose of a unitary matrix are both unitary.
4. The product of any two (or more) unitary matrices is unitary.
5. All properties of unitary matrices will be the properties of orthogonal matrices.

The class of unitary matrices is quite wide and it is possible to identify a number of unitary transform matrices that share common rules for their formation using fast transformation algorithms [8]. Included in these are the Fourier, Walsh, Haar, and Slant fast transforms. Much interest has been shown in recent years in presenting a unified view of these unitary transforms leading to the concept of a generalized transform to include all the common unitary transforms in a single class. This will be considered further in a later chapter.

3.3.3 Eigenvectors and eigenvalues

It is sometimes necessary to determine the similarity between matrices or, more specifically, to find alternative solutions to a matrix representation that are easier to calculate using the digital computer.

As an example consider an $N \times N$ matrix A operating on x_i which we would like to express in terms of a scalar λ instead of the matrix, i.e.

$$A . x_i = \lambda x_i. \tag{3.40}$$

A trivial solution is when $x = 0$. With a suitable choice of λ alternative solutions can often be found. Using the identity matrix we can write

$$(A - \lambda I)x_i = 0. \tag{3.41}$$

It can be shown that these solutions are realized if the determinant of the coefficients is zero [9].

$$\mathrm{Det}(A - \lambda I) = \begin{bmatrix} a_{11} - \lambda, \, a_{12}, & \dots, a_{1n} \\ a_{21}, & a_{22} - \lambda, \dots, a_{2n} \\ \vdots & \vdots & \vdots \\ a_{n1}, & a_{n2}, & \dots, a_{nn} - \lambda \end{bmatrix} = 0. \tag{3.42}$$

The roots of this algebraic equation are the special values of λ that we seek, denoted by λ_i. They are called the *eigenvalues* or *characteristic roots* of the matrix A. Corresponding to each eigenvalue there will be a solution of eqn (3.41) in the form of Cx_i, where C is an arbitrary constant. These solutions are called the *eigenvectors* or *characteristic vectors*.

A property of Hermitian (including real and symmetric) operators is that the eigenvectors corresponding to different eigenvalues are orthogonal [10]. This means that with a suitable scaling factor we can obtain an orthonormal set and so approximate other functions by superposition of a set of eigenfunctions in the same way as with a Fourier or other orthogonal series.

A development of this derivation of characteristic roots is to use the eigenvectors of a given matrix to express the matrix in terms of the product of a simple diagonal matrix and another matrix derived from the vector of interest [3].

Let matrix A have a distinct set of eigenvalues $\lambda_1, \lambda_2, \ldots, \lambda_N$ and let $x_1, x_2, \ldots, x_N$ be an associated vector set. If we define a matrix T consisting of x_i as columns and a second matrix T' formed by using x_i as rows, then

$$A = . \begin{bmatrix} \lambda_1 & & & 0 \\ & \lambda_2 & & \\ & & \ddots & \\ 0 & & & \lambda_b \end{bmatrix} . T'.$$

(3.43)

Here a diagonal matrix is formed from the eigenvalues of A arranged along the main diagonal of the matrix with zeros contained everywhere else. This is called *reduction to diagonal form* and plays a fundamental role in matrix theory and in the derivation of practical transform algorithms [11].

Eigenvalues are used directly in the derivation of the Karhunen–Loève series and the Karhunen–Loève transform (KLT) [12]. The KLT is of interest due to its extremely small residual MSE and for this reason is used as a standard of comparison for studying transformation techniques. It is not possible to derive a fast transform algorithm for use in digital computation, however, so the transform is little used for signal processing purposes.

3.3.4 Kronecker products

An important concept used in the derivation of a number of non-sinusoidal transforms is that of Kronecker products of matrices. In the previous section the matrix products referred to were all of equal size and of rectangular shape. Practical matrices we have to deal with may not necessarily be equal in size and a direct operator for matrix products which we can use in all cases is the *Kronecker product*, defined as $\otimes$.

If $M = [m_{ij}]$ is an $m \times p$ matrix and $N = [n_{ij}]$ is an $n \times q$ matrix then the

Kronecker product $M \otimes N$ is the $m . n \times p . q$ matrix given by

$$M \otimes N = \begin{bmatrix} m_{11}N, & m_{12}N, & \ldots, & m_{1p}N \\ m_{21}N, & m_{22}N, & \ldots, & m_{2p}N \\ \vdots & & \vdots & \\ m_{m1}N, & m_{m2}N, & \ldots, & m_{mp}N \end{bmatrix}. \tag{3.44}$$

An example will make this clear. Let

$$M = \begin{bmatrix} 1 & 1 \\ 1 & -1 \end{bmatrix} \quad \text{and} \quad N = \begin{bmatrix} -1 & 1 & 1 & 1 \\ 1 & -1 & 1 & 1 \\ 1 & 1 & -1 & 1 \\ 1 & 1 & 1 & -1 \end{bmatrix}$$

then the Kronecker product

$$M \otimes N = \begin{bmatrix} N & N \\ N & -N \end{bmatrix} = \begin{bmatrix} -1 & 1 & 1 & 1 & -1 & 1 & 1 & 1 \\ 1 & -1 & 1 & 1 & 1 & -1 & 1 & 1 \\ 1 & 1 & -1 & 1 & 1 & 1 & -1 & 1 \\ 1 & 1 & 1 & -1 & 1 & 1 & 1 & -1 \\ -1 & 1 & 1 & 1 & 1 & -1 & -1 & -1 \\ 1 & -1 & 1 & 1 & -1 & 1 & -1 & -1 \\ 1 & 1 & -1 & 1 & -1 & -1 & 1 & -1 \\ 1 & 1 & 1 & -1 & -1 & -1 & -1 & -1 \end{bmatrix}. \tag{3.45}$$

It is easy to show that the following relationships are valid for Kronecker products of matrices [3]

$$A \otimes B \otimes C = (A \otimes B) \otimes C = A \otimes (B \otimes C)$$
$$(A + B) \otimes (C + D) = A \otimes C + A \otimes D + B \otimes C + B \otimes D$$
$$(A \otimes B)(C \otimes D) = (A . C) \otimes (B . D) \tag{3.46}$$

where $A . C$ and $B . C$ are conventional matrix products.

Kronecker products of matrices are useful in factorizing the transform matrices and in developing number-theoretic, non-sinusoidal, and generalized transforms as we shall see later.

3.3.5 Matrix factorization

When we come to consider fast transforms in the next two chapters it will be convenient to deal with these as the result of matrix and vector products, since this can then lead to a way of expressing the transformation in algorithmic form and hence to a structured computer program.

The set of orthogonal transformations considered here can all be expressed as

$$A_N = B . X_N \qquad (3.47)$$

where X_N is the data series to be transformed, i.e. $X_N = x_i^{\mathsf{T}}$ ($i = 0, 1, \ldots, N-1$) and T is a transpose. The resulting transformed value A_N is also a transposed vector, $A_N = x_n^{\mathsf{T}}$. ($n = 0, 1, \ldots, N-1$). B is an $N \times N$ matrix appropriate to the transformation being carried out. For a Fourier transform for example B contains elements that are powers of $W = \exp(-2\pi \mathrm{j}/N)$ and which have an absolute value of 1.

The essence of a fast transformation algorithm lies in expressing B in terms of the product of a number of sparse matrices, also of size $N \times N$, which are simpler to evaluate than the composite matrix B. Thus we need a method of factorizing B into these simpler matrices [13]. Good has shown that matrices that are non-prime and composite can be factored into a number of highly redundant or sparse matrices, and the following gives an outline of his procedure. Proofs are not given and can be found in [13–15].

Factorization can only be attempted if N is *not* a prime number but is relevant if N can be decomposed into the product of other prime numbers, $N = N_1, N_2, \ldots, N_m$, where N_1, N_2 etc. are m prime number factors of N. Consider for example a matrix derived by the process of Kronecker multiplication. For a Kronecker matrix, K consisting of $N \times N$ entries $k_{i,n}$ where $i, n = 0, 1, \ldots, N-1$, there exist for the case of $N = 2^p p$ identical matrices, each of size $N \times N$ such that

$$K = \prod_{k=1}^{p} K_k. \qquad (3.48)$$

Each of these matrices is defined by

$$\begin{bmatrix}
k_{r,0,0} \ldots k_{r,0,p-1} & & & \\
& k_{r,0,0} \ldots k_{r,0,p-1} & & \\
& & \ddots & \\
& & & k_{r,0,0} \ldots k_{r,0,p-1} \\
k_{r,1,0} \ldots k_{r,1,p-1} & & & \\
& k_{r,1,0} \ldots k_{r,1,p-1} & & \\
& & \ddots & \\
& & & k_{r,1,0} \ldots k_{r,1,p-1} \\
\vdots \qquad \vdots & & & \\
k_{r,p-1,0} \ldots k_{r,p-1,p-1} & \vdots \qquad \vdots & & \\
& k_{r,p-1,0} \ldots k_{r,p-1,p-1} & & \\
& & \ddots & \\
& & & k_{r,p-1,0} \ldots k_{r,p-1,p-1}
\end{bmatrix}$$

$$(3.49)$$

where K_k has 2^{p+1} non-zero entries and all other matrix positions are filled with zeros.

The general case for Fourier matrix factorization is given below. Here B consists of elements, $W_{in} = \exp(-2\pi j/N)$ and $i, n = 0, 1, \ldots, N-1$ represent the row and column numbers.

Let B be a 12×12 matrix written as

$$
B = \begin{bmatrix}
1 & 1 & 1 & 1 & 1 & 1 & 1 & & 1 \\
1 & W & W^2 & & \cdots & & & & W^{11} \\
1 & W^2 & W^4 & & \cdots & & & & W^{22} \\
\vdots & \vdots & \vdots & & & & & & \vdots \\
1 & W^{11} & W^{22} & & \cdots & & & & W^{121}
\end{bmatrix}. \tag{3.50}
$$

Here N can be factorized into $N = N_1 \cdot N_2$ with $N_1 = 3$ and $N_2 = 4$. We could expect in this way to reduce eqn (3.46) into two $N \times N$ matrices B_1 and B_2. Actually, as we shall see later, one or more permutation matrices will also be necessary. Matrix B_1 may be derived as follows [15].

The first N_1 rows of B are divided into three sections, each containing N_2 values, i.e.

$$
\begin{array}{cccc|cccc|cccc}
1 & 1 & 1 & 1 & 1 & 1 & 1 & 1 & 1 & 1 & 1 & 1 \\
\hline
1 & W & W^2 & W^3 & W^4 & W^5 & W^6 & W^7 & W^8 & W^9 & W^{10} & W^{11} \\
1 & W^2 & W^4 & W^6 & W^8 & W^{10} & W^{12} & W^{14} & W^{16} & W^{18} & W^{20} & W^{22}
\end{array} \tag{3.51}
$$

The values found in each of these sections are then displaced in the column direction to become the right diagonal of an $N_2 \times N_2$ (i.e. 4×4) sub-matrix. The juxtaposition of these sub-matrices forms the B_1 factored matrix, i.e.

$$
B_1 = \begin{bmatrix}
1 & & & & 1 & & & & 1 & & & \\
& 1 & & & & 1 & & & & 1 & & \\
& & 1 & & & & 1 & & & & 1 & \\
& & & 1 & & & & 1 & & & & 1 \\
1 & & & & W^4 & & & & W^8 & & & \\
& W & & & & W^5 & & & & W^9 & & \\
& & W^2 & & & & W^6 & & & & W^{10} & \\
& & & W^3 & & & & W^7 & & & & W^{11} \\
1 & & & & W^8 & & & & W^{16} & & & \\
& W^2 & & & & W^{10} & & & & W^{18} & & \\
& & W^4 & & & & W^{12} & & & & W^{20} & \\
& & & W^6 & & & & W^{14} & & & & W^{22}
\end{bmatrix}. \tag{3.5.2}
$$

All the vacant spaces in this $N \times N$ matrix are filled with zeros. Matrix B_2 is defined as a right diagonal matrix, also of order $N \times N$, which contains along its main diagonal N equal sub-matrices of size $N_2 \times N_2$. These sub-matrices have elements of the form $W^{N1.ab}$, where $a, b = 0, 1 \ldots N_2$ and represent the row and column number.

Since $W^{n1} = \exp(-2\pi j/12)3 = -j$ we can write for B_2:

$$B_2 = \begin{bmatrix}
1 & 1 & 1 & 1 & & & & & & & & \\
1 & -j & -1 & j & & & & & & & & \\
1 & -1 & 1 & -1 & & & & & & & & \\
1 & j & -1 & -j & & & & & & & & \\
 & & & & 1 & 1 & 1 & 1 & & & & \\
 & & & & 1 & -j & -1 & j & & & & \\
 & & & & 1 & -1 & 1 & -1 & & & & \\
 & & & & 1 & j & -1 & -j & & & & \\
 & & & & & & & & 1 & 1 & 1 & 1 \\
 & & & & & & & & 1 & -j & -1 & j \\
 & & & & & & & & 1 & -1 & 1 & -1 \\
 & & & & & & & & 1 & j & -1 & -j
\end{bmatrix}$$

(3.53)

Both B_1 and B_2 are sparse matrices and it can be shown that calculation of their product requires less arithmetic operations than the calculation of the original matrix B despite the need to include one or more permutation matrices [13].

One consequence of splitting up the calculation in this way is that the final values of A_N are not obtained in a straightforward sequential order and this is why we need the permutation matrices. If we attempt to form $B = B_1 . B_2$ we find that the matrix differs from eqn (3.50) through a permutation of its rows. In addition to factorizing into B_1 and B_2 we need to multiply B_1 by B_2 by a permutation matrix, P to arrive at A_n in the correct order. This is different to the behaviour of the Kronecker matrices, where no such permutation matrix is needed. An example of this technique applied to a radix-2 FFT will be given in Chapter 4.

3.4 Some specialized matrices

Certain specific matrices are particularly relevant to orthogonal transformation. One of these is the *Hadamard matrix*, used as a basis for a number of non-sinusoidal orthogonal series, particularly the Walsh series. Another is the *Dyadic matrix*, which can be used for the same purpose. A number of *circulant matrices* also find a use in discrete Fourier transform expansions and play an important part in the derivation of the number-theoretic transforms considered later.

3.4.1 The Hadamard matrix

A Hadamard matrix H is an orthogonal matrix with elements having values of $+1$ and -1 only. The first two orders are

$$H_1 = [+1]$$

and

$$H_2 = \begin{bmatrix} +1 & +1 \\ +1 & -1 \end{bmatrix}. \tag{3.54}$$

The order of a Hadamard matrix is 1, 2 or $4n$, where n is an integer. Matrices of higher order than 4 are obtained through Kronecker multiplication. Hence,

$$H_4 = H_2 \otimes H_2 = \begin{bmatrix} H_2 & H_2 \\ H_2 & -H_2 \end{bmatrix} = \begin{bmatrix} +1 & +1 & +1 & +1 \\ +1 & -1 & +1 & -1 \\ +1 & +1 & -1 & -1 \\ +1 & -1 & -1 & +1 \end{bmatrix}. \tag{3.55}$$

In general

$$H_N = H_{N/2} \otimes H_2. \tag{3.56}$$

Hadamard matrices of the same order can be changed into other Hadamard matrices by permuting rows and columns and by multiplying rows and columns by -1. Not all Hadamard matrices of the same order can be changed in this way. However, those that have every element in their first row and column as $+1$ can always be changed and these are called *normalized Hadamard matrices*.

Rules for multiplication of Hadamard matrices are

$$H . H^T = N . I_N$$
$$H . H^T = H^T . H \tag{3.57}$$

where N is the order of the matrix.

Other properties follow from those of the Kronecker product given previously.

3.4.2 The Dyadic matrix

The *Dyadic matrix* is formed by the addition modulo-2 of two numbers, shown as R and S in Table 3.1. The value of this matrix for transform evaluation is that the eigenvalues of any dyadic matrix express a set of discrete Walsh functions [16]. This gives a relationship with the

Table 3.1 Modulo-2 additions for $N = 16$

	R															
S	1	2	3	4	5	6	7	8	9	10	11	12	13	14	15	16
1	0	3	2	5	4	7	6	9	8	11	10	13	12	15	14	17
2	3	0	1	6	7	4	5	10	11	8	9	14	15	12	13	18
3	2	1	0	7	6	5	4	11	10	9	8	15	14	13	12	19
4	5	6	7	0	1	2	3	12	13	14	15	8	9	10	11	20
5	4	7	6	1	0	3	2	13	12	15	14	9	8	11	10	21
6	7	4	5	2	3	0	1	14	15	12	13	10	11	8	9	22
7	6	5	4	3	2	1	0	15	14	13	12	11	10	9	8	23
8	9	10	11	12	13	14	15	0	1	2	3	4	5	6	7	24
9	8	11	10	13	12	15	14	1	0	3	2	5	4	7	6	25
10	11	8	9	14	15	12	13	2	3	0	1	6	7	4	5	26
11	10	9	8	15	14	13	12	3	2	1	0	7	6	5	4	27
12	13	14	15	8	9	10	11	4	5	6	7	0	1	2	3	28
13	12	15	14	9	8	11	10	5	4	7	6	1	0	3	2	29
14	15	12	13	10	11	8	9	6	7	4	5	2	3	0	1	30
15	14	13	12	11	10	9	8	7	6	5	4	3	2	1	0	31
16	17	18	19	20	21	22	23	24	25	26	27	28	29	30	31	0

Walsh function that is used to derive an invariant form of the Walsh spectrum [16] and also finds a place in logic analysis [17]. Like the Walsh and several other non-sinusoidal matrices, the dyadic matrix forms its own inverse, which can be deduced by observation of the symmetry about its diagonal in Table 3.1. Products of dyadic matrices result in a further dyadic matrix, which is commutative.

3.4.3 Circulant matrices

A *Circulant matrix* is a square matrix in which the elements in each row are obtained by a circular right-shift of the elements in the preceding row. An example is

$$C_4 = \begin{bmatrix} c_{11} & c_{12} & c_{13} & c_{14} \\ c_{14} & c_{11} & c_{12} & c_{13} \\ c_{13} & c_{14} & c_{11} & c_{12} \\ c_{12} & c_{13} & c_{14} & c_{11} \end{bmatrix}. \tag{3.58}$$

The circulant matrix may be reduced to diagonal form through the discrete Fourier transform matrix. The diagonal elements are then the Fourier series expansion of the elements in the first row of the circulant matrix. Two further properties are that the circulant matrix forms its own inverse and that a product formed from circulant matrices will form a circulant matrix that is commutative.

An interesting circulant matrix which we will meet later in transform theory is the *block circulant matrix*. This is a square matrix whose sub-matrices are individually circulant. Here the sub-matrices in any row can be obtained by a right circular shift of the preceding row of sub-matrices. An example is

$$B_4 = \begin{bmatrix} C_0 & C_1 & C_2 & C_3 \\ C_3 & C_0 & C_1 & C_2 \\ C_2 & C_3 & C_0 & C_1 \end{bmatrix} \tag{3.59}$$

where each sub-matrix C_j $(j = 0, 1, \ldots, N-1)$ is a circulant matrix, such as C_4 in eqn (3.58).

3.4.4   Circular convolution

An example of a circulant matrix is found in the matrix representation of *circular convolution* (Section 1.4.1). This is the convolution of signals that repeat at intervals defined by the discrete sample length N, so that from eqn (1.24)

$$y = \sum_{i=0}^{N-1} x_i h(\tau - i) \tag{3.60}$$

$(i, \tau = 0, 1, \ldots, N-1$, neglecting scaling by $1/N$). $h(i)$ is by definition periodic in N terms, so that

$$h(\tau - i) = h(\tau - i + N). \tag{3.61}$$

Equation (3.60) can be expressed in matrix vector form as

$$[y] = H[x] \tag{3.62}$$

where H is a circulant matrix as C_4 in eqn (3.58).

3.5   Discrete data manipulation

A number of mathematical techniques are encountered that apply to the handling of discrete sampled data. Some of the most common of these are described in this section.

3.5.1   Bit reversal

A sequence of numbers in ascending natural order can be rearranged such that the digital coefficients representing the numbers are assembled in reverse order. This gives a sequence in decimal order referred to as *bit-reversed notation* or bit-reversed order (we will see later why we need to do this).

Let an integer D be represented as an m-bit binary number B, e.g.

$$(D)_{10} = (b_{m-1}2^{m-1}, +b_{m-2}2^{m-2}, \ldots, +b_2 2^2 + b_1 2^1 + b_0 2^0)$$

so that

$$B = (b_{m-1}, b_{m-2}, \ldots, b_2, b_1, b_0) \qquad (3.63)$$

where b = logical 0 or 1. The bit-reversal operation will result in

$$(D)'_{10} = (b_0 2^{m-1} + b_1 2^{m-2} +, \ldots, + b_{m-2}2^1 + b_{m-1}2^0)$$

so that

$$B' = (b_0, b_1, b_2, \ldots, b_{m-2} \cdot b_{m-1})_2. \qquad (3.64)$$

Thus if $(D)_{10} = 3 = (011)_2$, then bit-reversal gives $(110)_2 = 6 = (D')_{10}$.

A bit-reversal table for $N = 8$ is given in Table 3.2. Algorithms to carry out this operation are discussed later in this book.

Table 3.2 Bit-reversal table for $N = 8$

Decimal number	Binary representation	Bit-reversal	Decimal number representation
0	000	000	0
1	001	100	4
2	010	010	2
3	011	110	6
4	100	001	1
5	101	101	5
6	110	011	3
7	111	111	7

3.5.2 Modulo-2 arithmetic

Modulo-2 addition is equivalent to an EXCLUSIVE-OR operation, defined as

$$0 \oplus 0 = 0 \qquad 0 \oplus 1 = 1 \qquad 1 \oplus 0 = 1 \qquad 1 \oplus 1 = 0 \qquad (3.65)$$

where $\oplus$ denotes addition without carry to the next digit. Thus $1010 \oplus 1110 = 0100$.

Extending this definition to negative numbers, we write

$$(-a) \oplus (-b) = (a \oplus b)$$
$$(-a) \oplus (+b) = -(a \oplus b)$$
$$(+a) \oplus (-b) = -(a \oplus b) \qquad (3.66)$$

where a, b are either logical 0 or 1.

A table for modulo-2 addition was given to illustrate the Dyadic matrix in Table 3.1. Modulo-2 multiplication is the Kronecker multiplication considered earlier (Section 3.3.3).

3.5.3   Gray code conversion

Translation of a decimal number into the Gray code equivalent is carried out by first transforming the decimal number into a binary representation. Modulo-2 addition is then performed on each bit with its immediate neighbour on the left. Translation of this result into decimal notation then gives the Gray code equivalent. Thus 11_{10} becomes 1011_2 and modulo-2 addition of pairs of bits gives 1110_2 or 14_{10}.

Table 3.3   Gray code conversion table

Decimal	Code	Decimal	Code
0	0000	8	1100
1	0001	9	1101
2	0011	10	1111
3	0010	11	1110
4	0110	12	1010
5	0111	13	1011
6	0101	14	1001
7	0100	15	1000

This may be expressed in terms of a binary bit string as follows. A binary number may be expressed as

$$B = (b_m, \, _{m-1}, \ldots, b_0)_2 \tag{3.67}$$

where b_i gives the position in the binary number. This is given in the Gray code as

$$B_g = (g_m, g_{m-1}, \ldots, g_0)_2 \tag{3.68}$$

where $g_i = b_i + b_{i+1}$.

A Gray code conversion table is given in Table 3.3 for the first 16 decimal numbers.

References

1. Birkhoff, G. and Maclave, S. *A survey of modern algebra*. Macmillan, New York (1977).
2. Stoll, R. R. *Linear algebra and matrix theory*. Dover, New York (1952).
3. Bellman, R. *Introduction to matrix analysis*. McGraw-Hill, New York (1960).
4. McClellan, J. H. *Number theory in digital signal processing*. Prentice-Hall, Englewood Cliffs (1979).

5. Beauchamp, K. G. and Yuen, C. K. *Digital methods for signal analysis.* George Allen & Unwin, London (1979).

6. Andrews, H. and Caspari, K. Orthogonal transformation. In *Computer techniques in image processing* (ed. H. C. Andrews) pp. 73–102. Academic Press, New York (1970).

7. Andrews, H. C. and Kane, J. Kronecker matrices, computer implementation and generalized spectra. *J. Ass. Comput. Mach.* **17** (2), 260–8 (1970).

8. Fino, B. J. and Algazi, V. R. A unified treatment of discrete fast unitary transforms. *SIAM J. Comput.* **6** (4), 700–17 (1977).

9. Noble, B. *Numerical methods.* Oliver & Boyd, Edinburgh (1964).

10. Wallis, J. S. *Hadamard matrices.* Lecture notes 292NY, Springer, Berlin (1972).

11. Dickinson, B. and Steiglitz, K. An approach to the diagonalization of the discrete Fourier transform. IEEE Conference on Acoustics, Speech and Signal Processing, pp. 227–30 (1980).

12. Maqusi, M. *Applied Walsh analysis.* Heyden, London (1981).

13. Good, I. J. The interaction algorithm and practical Fourier series. *J. R. statist. Soc.* **B20,** 361–72 (1958); **B22,** 372–5 (1960).

14. Kahaner, D. K. Matrix description of the fast Fourier transform. *IEEE Trans. Audio Electroacoust.* **AU-18,** (4), 442–51 (1970).

15. Theilheimer, F. A matrix version of the fast Fourier transform. *IEEE Trans. Audio Electroacoust.* **AU-17,** (2), 158–61 (1969).

16. Kak, S. C. On matrices with Walsh functions of the eigenvectors. Proceedings of the Symposium on Applied Walsh Functions, Washington DC, pp. 384–7. AD 744650. Available from Natl. Technical Info. Service, US Dept. Commerce, Springfield, Virginia 22151 USA.

17. Hurst, S. L., Miller, D. M., and Muzio, J. C. *Spectral techniques in digital logic.* Academic Press, New York (1984).

4

FAST FOURIER TRANSFORMATIONS

4.1 Introduction

The symmetries noted in Chapter 2 as being associated with the DFT and IDFT may be exploited to carry out transformation efficiently with significant savings in computational complexity through the fast Fourier transform algorithms. Similar symmetries found in other orthogonal transformations, e.g. Walsh, Haar, cosine, etc. can also lead to the development of fast transformation techniques. The Karhunan–Loève transform, although providing vector coefficients having the maximum energy compaction and therefore optimum transform efficiency [1], does not lend itself to fast algorithmic techniques, so only the *sub-optimum* transformations will be discussed here.

The advantages of the fast transform techniques for digital calculation are well-known. Their disadvantages are not so frequently discussed and merit some consideration at this stage. First, the size of the database to be transformed may not justify the additional complexity of fast algorithms such as the data rearrangements required prior or subsequent to a transformation. Taking this and other factors into consideration, fast algorithmic techniques give no real advantage in computation for $N < 50$. Where N is less than this the simpler methods are to be preferred.

This rule does not apply to the special 'small-N' transforms used as building blocks for the number-theoretic transforms considered in Section 4.3.2. Here N is nearly always less than 16, but these special transforms are used as a particular way of partitioning a large number for N and not as transforms in their own right.

A second disadvantage is that except for special hardware parallel-processing devices, the fast Fourier transform (FFT) is not performed in 'real time'—we need to have all N values of x together at the beginning of the FFT process. We are also somewhat restricted in the value of N. The most economical arrangement is for N to be a power of two, but this may not always be possible for real data. Any technique we use for $N \neq$ power of two inevitably increases program complexity and this has to be considered against the speed advantages of the fast transform.

Finally the accuracy of the FFT may not provide what is required, particularly with regard to round-off noise and data scaling. Compared with non-FFT methods it may be necessary to increase the word length of the data to overcome this problem [2–4].

Having said all this, there remain overwhelming advantages in the use of fast transform techniques in very many situations. Their use has enabled many complex calculations to be performed which, in the absence of fast techniques, would not otherwise be possible even with the very efficient computing hardware available today.

The earliest algorithms used to carry out fast Fourier transform calculations were reported by Runge and Koenig in 1924 [5] and by Danielson and Lanczon in 1942 [6]. These early methods relied on the symmetries of the sine and cosine functions to achieve computational economy. The fast Fourier transform described Cooley and Tukey in 1965 went further and achieved a considerable reduction in the number of mathematical operations required [7]. It also provided a general method applicable when N is composite and not necessarily a power of two. A slightly earlier paper by Good [8] also described a fast transform method but required that the factors of N be mutually prime, which was at first seen as a major disadvantage for the method.

It was in fact the Cooley–Tukey paper that initiated the application of transform methods using the digital computer to a wide range of problems which, due to the computational effort involved, were not tractable without FFT methods.

The major advantage of the Cooley–Tukey method, and indeed of all subsequent fast transform algorithms, lies in the reduction in the number of arithmetic operations required. Applying the DFT eqn (2.35) to obtain a set of N transformed values requires about N^2 complex multiplications and N^2 complex additions. The FFT algorithms enable these figures to be reduced to $(N/2)\log_2(N/2)$ complex multiplications and $N \log_2 N$ complex additions. Actually, rather less complex multiplications would be required, as there are rows in the FFT matrix that require only a subtraction. This point will be considered later.

Since the publication of the Cooley–Tukey algorithm there has been considerable development in alternative algorithms that achieve similar or better computational efficiency, or that offer alternative organizational advantages. They are all called fast Fourier transforms and will be distinguished from each other by the originators' initials (e.g. C–T algorithm). The key to these methods lies in their exploitation of the possibilities for factorizing the number of values of the data series to be transformed. An explanation will be given first in terms of matrix factorization. This will be followed by the derivation of a general computing algorithm.

FFT algorithms are at their most efficient only when N is not a prime number, i.e.

$$N = r_1 r_2 r_3 \ldots r_m, \quad \text{and} \quad m > 1$$

where

$$r_1 = r_2 = r_3 = \ldots = r_m = r \quad \text{then} \quad N = r^m \qquad (4.1)$$

and we refer to these as fixed radix or radix-r transforms. Most algorithms in general use are designed for radix-2 operation. The C–T algorithm and the alternative Sande–Tukey (S–T) algorithm are of this type [9]. They differ in the ordering of the sequence of intermediate calculations and are referred to as *decimation in time* and *decimation in frequency* algorithms, respectively, for reasons that will become apparent later.

With radix-2 algorithms the constraint limiting N to 2^p, where p is an integer, can prove restrictive. A number of algorithms have been published using radix-4, radix-8 and even mixed-radix coefficients [10–12]. These can give savings in software or hardware implementations in terms of multiplications required, but the savings are very dependent on the size of N. FFT algorithms for arbitrary factors have also been proposed for $N = r_1 r_2 \ldots r_m$, where r_i is an integer [13].

A feature of the DFT that gives rise to some economies in the design of an FFT is its complex conjugate property (see Section 2.3.1). Since a real transformation of length N has conjugate symmetry, $X_{N-n} = X_n^*$, then half the transformation from a signal x_i is redundant and need not be calculated. Several examples of the use of this property are given later.

In recent years some new developments in fundamental transform theory have given rise to several algorithms that are more efficient than the C–T and S–T algorithms in terms of the number of arithmetic operations required. These are the *number-theoretic* transforms of Winograd, McClellan, Nussbaumer and others [14–17]. They lead to fairly complex and lengthy computer algorithms but can be considerably faster then the C–T-based algorithms. Their principal value is likely to be in applications using very large data sequences, such as those found in image processing or in real-time operations using parallel processing techniques.

Other alternative methods of fast Fourier transformation have been described that fall outside this group of number-theoretic transforms. One of these is the *chirp-Z transform* which is used in certain microelectronics hardware developments. This will be described later in (Chapter 6) when the hardware FFT is considered.

4.2 Fast Fourier transform algorithms

The FFT is an algorithm for computing the DFT of a series in a way that requires less computational effort than the direct calculation. Many versions of the algorithm have been published (see [1–13]) and are available as subroutines in mathematical software packages [18–20], so there is usually little need for the reader to write his own. The only possible exception is for microprocessor or hardware design, where machine-dependent software is required. It is useful however to understand the mathematical working and characteristics of the various FFT algorithms and techniques available so as to be able to use them effectively in a given situation. The treatment of the FFT and other fast algorithms that follows is therefore presented in a heuristic manner and is not intended to be either complete or rigorous. Adequate references are included to enable the reader to pursue proofs and derivations where these are required.

The DFT as considered earlier is a $1:1$ conversion of any sequence of numbers x_i ($i = 0, 1, 2, \ldots, N-1$), consisting of N complex numbers, into another sequence defined by

$$X_n = \sum_{i=0}^{N-1} x_i W^{in} \qquad (4.2)$$

($i, n = 0, 1, 2, \ldots, N-1$) where $W = \exp(-j2\pi/N)$.

The factor $1/N$ will be omitted in this and later derivations in order to simplify notation. It serves as a scaling factor to the values of X_n obtained and needs to be considered only when numerical values are to be attributed to the frequency (or other) scale.

The direct calculation of eqn (4.2) requires the solution of a matrix of terms $[X_n] = [W^{in}][x_i]$, which are all complex. For example where $N = 8$, the matrix takes the form of Fig. 4.1. If matrix multiplication is carried out in this way, we must calculate 64 complex products and 64 complex additions. In the general case N^2 complex multiplications and additions are required.

The C–T algorithm on the other hand is a recursive algorithm based on matrix factorization, as discussed in the previous chapter. It reduces the matrix to a series of sparse matrices (i.e. simpler matrices having many zero terms). Using this technique much of the redundancy in the

$$
\begin{bmatrix} X_0 \\ X_1 \\ X_2 \\ X_3 \\ X_4 \\ X_5 \\ X_6 \\ X_7 \end{bmatrix}
=
\begin{bmatrix}
W^0 & W^0 & W^0 & W^0 & W^0 & W^0 & W^0 & W^0 \\
W^0 & W^1 & W^2 & W^3 & W^4 & W^5 & W^6 & W^7 \\
W^0 & W^2 & W^4 & W^6 & W^8 & W^{10} & W^{12} & W^{14} \\
W^0 & W^3 & W^6 & W^9 & W^{12} & W^{15} & W^{18} & W^{21} \\
W^0 & W^4 & W^8 & W^{12} & W^{16} & W^{20} & W^{24} & W^{28} \\
W^0 & W^5 & W^{10} & W^{15} & W^{20} & W^{25} & W^{30} & W^{35} \\
W^0 & W^6 & W^{12} & W^{18} & W^{24} & W^{30} & W^{36} & W^{42} \\
W^0 & W^7 & W^{14} & W^{21} & W^{28} & W^{35} & W^{42} & W^{49}
\end{bmatrix}
\bullet
\begin{bmatrix} x_0 \\ x_1 \\ x_2 \\ x_3 \\ x_4 \\ x_5 \\ x_6 \\ x_7 \end{bmatrix}
$$

Fig. 4.1 Matrix description of DFT derivation.

calculation of repeated products required by eqn (4.2) can be avoided enabling a large reduction in calculation time to be realized.

Let us consider first a matrix consisting of r rows and s columns, in which the total number of points is $N = rs$. A Fourier transform of the elements of this matrix can then be arranged by carrying out s parallel Fourier transforms, each of r data points, on the individual columns and then adding the results. We have thus reduced the problem to a summation of a number of much shorter Fourier transforms instead of the calculation of one long transform. Quite apart from the saving in time (which will be shown below), this method has another major advantage in that a much smaller storage space is required to hold the intermediate calculations.

To illustrate the working of the algorithm in matrix terms a simple two-level factorizing for the total number of terms, $N = rs$, will first be considered.† Using this assumption we can express the N variables, n and i in eqn (4.2), as four sub-series for i_0, i_1, n_0, and n_1

$$n = n_1 r + n_0 \tag{4.3}$$

and

$$i = i_1 s + i_0 \tag{4.4}$$

where

$n_0 = 0, 1, 2, \ldots, r - 1;$

$n_1 = 0, 1, 2, \ldots, s - 1;$

$i_0 = 0, 1, 2, \ldots, s - 1;$

$i_1 = 0, 1, 2, \ldots, r - 1.$

We can see that i and n each still contain N discrete values if we take some particular values and substitute these in the above. For example if $N = 20 = rs$, where $r = 5$ and $s = 4$, then n_0 goes from 0 to 4, n_1 goes from 0 to 3, i_0 goes from 0 to 3, and i_1 goes from 0 to 4. Table 4.1 for n shows the inclusion of all 20 values as required. Similar values for i are given in Table 4.2.

Equation (4.2) can now be rearranged for this example as two sums,

$$X_{n1.n0} = \sum_{i_0=0}^{i_0=s-1} \sum_{i_1=0}^{i_1=r-1} x_{i1.i0} W^{in}. \tag{4.5}$$

Equation 4.5 is a recursive formula in which we carry out complex multiplications and additions at every sth sample of the $N - 1$ possible variables for i within the inner loop, and then advance the starting point

---

† This, and a later description of radix-2 transformation, are included by permission of George Allen & Unwin Ltd, from the authors previous publication (with C. K. Yuen) entitled, 'Digital Methods for Signal Analysis' (1979).

Table 4.1 Factorized values of n				
$n_1 =$	0	1	2	3	n_0
$n =$	0	5	10	15	0
	1	6	11	16	1
	2	7	12	17	2
	3	8	13	18	3
	4	9	14	19	4
	$r = 5$

Table 4.2 Factorized values of i					
$i_1 =$	0	1	2	3	4	i_0
$i =$	0	4	8	12	16	0
	1	5	9	13	17	1
	2	6	10	14	18	2
	3	7	11	15	19	3
	$s = 4$

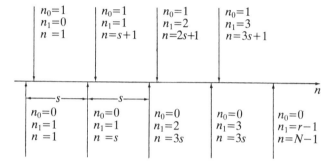

Fig. 4.2 Fast Fourier transform—simple partitioning.

for the first value of i in the outer loop. The inner loop procedure is then repeated from this new commencing point (see Fig. 4.2).

If we now expand W^{in} using eqns (4.3) and (4.4) we have

$$W^{in} = W^{(i_1 s + i_0)(n_1 r + n_0)}$$

$$= W^{i_1 n_1 rs} \cdot W^{n_0 i_1 s} \cdot W^{i_0(n_0 + n_1 r)} \qquad (4.6)$$

but $i_1 n_1 rs = i_1 n_1 N$ and, since i_1, n_1 is an integer by definition, the exponent is a multiple of 2π and

$$W^{i_1 n_1 N} = \exp\left(-j\frac{2\pi}{N}\right)^{i_1 n_1 N} = 1. \qquad (4.7)$$

We may also note for later use that

$$W^{N/2} = \exp(-j2\pi/2) = -1. \qquad (4.8)$$

Equation (4.5) now becomes

$$X_{(n_1 n_0)} = \sum_{i_0=0}^{i_0=s-1} \sum_{i_1=0}^{i_1=r-1} X_{(i_1, i_0)} W^{n_0 i_1 s} \cdot W^{i_0(n_0 + n_1 r)}. \qquad (4.9)$$

The inner sum over i_1 depends only on i_0 and n_0 and can be defined as a new series

$$X_{(i_0,n_0)} = \sum_{i_1=0}^{i_1=r-1} X_{(i_1,i_0)} W^{n_0 i_1 s} \qquad (4.10)$$

which we can recognize as the Fourier transform of a reduced array

$$x_{(i_1,i_0)}$$

Equation (4.9) can now be rewritten as

$$X_{(n_1,n_0)} = \sum_{i_1=0}^{i_1=s-1} x_{1(i_0,n_0)} W^{i_0(n_0+n_1 r)}. \qquad (4.11)$$

There are N elements in array x_1, but unlike the original array x_i we only require r operations to evaluate the transform, giving a total of Nr operations for array x_1 (remembering that these are complex operations). Similarly, we require Ns operations in order to calculate x_1 from x_i. Consequently this simple two-step algorithm requires a total of $T = N(r+s)$ complex operations. We can see how this procedure may be extended by increasing the number of steps to p, when the number of operations will be $T' = N(s_1 + s_2 + , \ldots, s_p)$, where $N = s_1, s_2, \ldots, s_p$. If $s_1 = s_2 = s_3, \ldots, = s_p = s$ then $N = s^p$, so that s can be expressed as the radix of the number series and we can write $p = \log_s N$, from which

$$T' = sN \log_s N = N\left(\frac{s}{\log_2 s}\right)\log_2 N. \qquad (4.12)$$

A plot of s against $(s/\log_2 s)$ indicates that radix $s = 3$ is the most efficient, and while algorithms for radix-3 have been proposed [21], very little loss of efficiency occurs if powers-of-2 radices are chosen.

Using radix-2 an improvement in computing speed over direct evaluation is obtained as

$$N^2/(2N \log_2 N) = N/(2 \log_2 N). \qquad (4.13)$$

4.2.1 Radix-2 transforms

The implementation of the FFT as described by Cooley and Tukey required that N be a power of two. This algorithm will be developed below by considering it as a general recursive device programmed by the digital computer. The original series is converted in stages, each stage providing an intermediate transform that supplies the input for subsequent stages, until the final series becomes the discrete transform of the original series.

With N limited to a power of two it is possible to express i and n in terms of the index p as a *binary-weighted series*

$$i = i_{p-1}2^{p-1} + i_{p-2}2^{p-2} + , \ldots , + i_1 2 + i_0 \qquad (4.14)$$

$$n = n_{p-1}2^{p-1} + n_{p-2}2^{p-2} + , \ldots , + n_1 2 + n_0. \qquad (4.15)$$

In this way we can express all the N possible values of the indices in terms of a binary number and hence facilitate the consideration of storage in the digital computer. The input and output series of the Fourier transform, x_i and X_n, are considered as being stored within the memory at location addresses defined by the binary representation of i and n. Using this convention, eqn (4.2) can be written in terms of a factorized sum, exactly as carried out previously for $N = rs$ but with the factorizing expressed in a binary-weighting form, i.e.

$$X_{(n_{p-1}, n_{p-2}, \ldots, n_0)} = \sum_{i_0=0}^{i_0=1} \sum_{i_1=0}^{i_1=1}, \ldots, \sum_{i_{p-1}=0}^{i_{p-1}=1}$$

$$x_{(i_{p-1}, i_{p-2}, \ldots, i_0)} \cdot W^{n(i_{p-1}2^{p-1} + i_{p-2}2^{p-2} +, \ldots, i_0)}. \qquad (4.16)$$

That this is equivalent to eqn (4.2) can be seen if we consider a simple case of

$$N = 2^2 = 4_{10} = \underset{i_2}{\diagup} \; 1 \; \underset{i_1}{\overset{|}{0}} \; \underset{i_0}{\diagdown} 0_2$$

In the decimal case

$$\sum_{i=0}^{i=4} x_{i_{(10)}} = x_0 + x_1 + x_2 + x_3$$

and in the binary case

$$\sum_{i_0=0}^{1} \sum_{i_1=0_2}^{1} \sum_{i=0}^{1} x_{i_{(2)}} = x_{000} + x_{001} + x_{010} + x_{011} + x_{100}$$

which is seen to contain all the possible values of x_i in the range up to decimal 4, expressed in binary form.

If we now expand the first term of the exponential for eqn (4.16) as we did in eqn (4.6)

$$W^{n i_{p-1} 2^{p-1}} = W^{n_{p-1} 2^{p-1} i_{p-1} 2^{p-1}} \cdot W^{n_{p-2} 2^{p-2} i_{p-1} 2^{p-1}}, \ldots, W^{n_1 2 i_{p-1} 2^{p-1}} \cdot W^{n_0 i_{p-1} 2^{p-1}} \qquad (4.17)$$

we find that all the intermediate terms go to unity and

$$W^{n i_{p-1} 2^{p-1}} = W^{n_0 i_{p-1} 2^{p-1}}. \qquad (4.18)$$

This enables the innermost sum of eqn (4.17) over $i_p - 1$ to be written as a shorter Fourier transform.

$$x_{1(n_0, i_{p-2}, \ldots, i_0)} = \sum_{i_{p-1}=0}^{i_{p-1}=1} x_{(i_{p-1}, i_{p-2}, \ldots, i_0)} W^{n_0 i_{p-1} 2^{p-1}} \qquad (4.19)$$

which depends only on $n_0, i_{p-2}, \ldots, i_0$.

Unlike the complete Fourier transform, this sum consists of a set of N numbers only, each calculated from two of the original data points. Subsequent sums, proceeding outwards in eqn (4.16), can be calculated using a generalized recursive expression for the exponential term

$$W^{ni_{p-q}2^{p-q}} = W^{(n_{q-1}2^{q-1}+n_{q-2}2^{q-2}+,\ldots,n_0)i_{p-q}2_{p-q}} \tag{4.20}$$

($q = 1, 2, 3, \ldots, p$). The successive sums are evaluated according to the equation

$$x_{q(n_0,n_1,\ldots,n_{q-1},i_{p-q-1},i_{p-q-2},\ldots,i_0)}$$
$$= \sum_{i_{p-q}=0}^{i_{p-q}=1} x_{q-1(n_0,n_1,\ldots,n_{q-2},i_{p-q},\ldots,i_0)}$$
$$W^{(n_{q+1}2^{q-1}+n_{q-2}2^{q-2}+,\ldots,n_0)i_{p-q}2^{p-q}} \tag{4.21}$$

($q = 1, 2, 3, \ldots, p$), which is the definition of the C–T algorithm.

To apply this recursive formula, the initial set of data x_i is first made equal to x_0 ($q = 1$); thus

$$x_i = x_{(i_{p-1},i_{p-2},\ldots,i_0)} = x_{0(i_{p-1},i_{p-2},\ldots,i_0)}.$$

This leads to the derivation of succeeding arrays in x_q so that the final array will be that for X_n. Since we have represented the two sets of p arguments for x_q, namely $(n_0, \ldots, n_{q-1}, i_{p-q-1}, \ldots, i_0)$, as binary representations of their storage locations, this can be used to simplify the working of the algorithm and so reduce the storage requirements for the intermediate sums. This involves fetching values from these two storage locations, carrying out complex multiplication and addition in accordance with eqn (4.2) and putting the results back into these same two locations. Over-writing of the input array with output values thus occurs. This indexing scheme has one disadvantage in that the elements of the final array in X_n are stored in the incorrect order in the core memory. Thus for the last array calculated

$$X_{(n_{p-1},n_{p-2},\ldots,n_0)} = x_{p(n_0,\ldots,n_{p-2},n_{p-1})}. \tag{4.22}$$

This shows that in order to find X_n the order of the bits in the binary representation of n must be reversed so as to obtain the index of the memory location where X_n may be found from the x_p array. From eqn (4.21) we see that for each element of x_q, one complex multiply/add operation is required. Also, a further complex multiply/add operation is required for the recursive generation of the complex exponential. Thus the total number of operations is $T' = 2Np$. But, since there are $N = 2^p$ elements in each array x_q, where $q = 1, 2, 3, \ldots, p$, the improvement in speed over direct evaluation is $T/T' = N/(2 \log_2 N)$, which is the result obtained earlier for the general case.

An alternative form of the algorithm is obtained if the roles of the indices i and n are interchanged when carrying out the expansion of the exponential term in eqn (4.17). This is known as the Sande–Tukey (S–T) version [9] and leads to the following recursion

$$
x'_q(n_0, n_1, \ldots, n_{q-1}, i_{p-q-1}, i_{p-q-2}, \ldots, i_0)
$$

$$
= \sum_{i_{p-q}=0}^{i_{p-q}=1} x_{q-1}(n_0, n_1, \ldots, n_{q-2}, i_{p-q}, \ldots, i_0)
$$

$$
\cdot W^{(i_{p-q}2^{p-q}+, \ldots, i_0)n_{q-1}2^{q-1}} \tag{4.23}
$$

which is the definition of the S–T algorithm. The two algorithms are operationally very similar although some advantages in speed, particularly for in-place array manipulation, are claimed for the S–T algorithm due to its simpler exponential structure.

Another way of considering these algorithms which shows rather more clearly their recursive nature is the *successive doubling method* described below.

Referring again to eqn (4.2) if we consider the series x_i to be divided into two interleaved series,

$$
x_{2i} = y_i
$$

$$
x_{2i+1} = z_i \tag{4.24}
$$

$(i = 0, 1, 2, \ldots, (N/2 - 1)$, where y_i consists of the even-numbered samples and z_i the odd-numbered samples, the discrete Fourier transforms of these two series can now be written (omitting the scaling factor $2/N$) as

$$
Y_n = \sum_{i=0}^{(N/2)-1} y_i W^{2in} \tag{4.25}
$$

$$
Z_n = \sum_{i=0}^{(N/2)-1} z_i W^{2in}. \tag{4.26}
$$

Note that i, n now represents a half-length series so that $W_{N/2} = \exp[(-2\pi j)/(N/2)] = W_N^2$. The transformed series Y_n and Z_n are displaced by one sampling interval so that to obtain the DFT from the complete N-point series using (4.25) and (4.26), we write

$$
X_n = \sum_{i=0}^{(N/2)-1} [y_i W^{n(2i)} + z_i W^{n(2i+1)}]
$$

$$
= \sum_{i=0}^{(N/2)-1} y_i W^{2in} + W^n \sum_{i=0}^{(N/2)-1} z_i W^{2in} \tag{4.27}
$$

since W^n is a constant for a given value of n. Hence

$$
X_n = Y_n + W^n Z_n. \tag{4.28}
$$

But n is limited to $(N/2) - 1$ samples, so that to complete the sequence for X_n we need to find the terms from $N/2$ to $N - 1$. We note from the theory of the DFT that for $n > N/2$ the transforms Y_n and Z_n will repeat periodically the values obtained with $0 < n \leqslant N/2$, so that $(n + N/2)$ can be substituted for n in the phase shift term only, i.e.

$$X_{(n+N/2)} = Y_n + W^{n+N/2}Z_n. \tag{4.29}$$

But

$$W^{n+N/2} = \exp(-j2\pi n/N - j\pi),$$

which from eqn (4.8)

$$= -\exp(-j2\pi n/N) = -W^n. \tag{4.30}$$

Hence

$$X_{(n+N/2)} = Y_n - W^n Z_n. \tag{4.31}$$

Equations (4.28) and (4.31) enable the first and last $N/2$ points to be evaluated from the separate transforms formed from $x_{(2i)}$ and $x_{(2i+1)}$. Neglecting addition and subtraction as taking negligible computing time compared with multiplication, we see that this result requires $N + 2(N/2)^2$ multiplications compared with N^2 for direct evaluation. For large values of N this represents a reduction of the computing time required by almost half.

Equations (4.28) and (4.31) describe the basic operations of the radix-2 FFT and, as we shall see later, are normally carried out together as a single operation known as a 'butterfly' computation, with W^n as the complex operator (often called the *twiddle factor*).

The process of dividing the series by two and then obtaining the discrete Fourier transform from the transforms of each half of the series is repeated to obtain further reductions in computing time. In the limit, providing that the series is capable of division by two, a two-point transform is obtained from two single-point transforms. The DFT of a single point is, however, the point itself. Thus

$$X_n = \sum_{i=0}^{i=0} x_i W^{in} = x_i \tag{4.32}$$

$(i = 0)$, so that the method reduces to a simple series of additions and multiplications of the twiddle factors W^n. In this limiting case the improvement in computing speed is $N/(2 \log_2 N)$, as obtained previously. This approach to the evaluation of the C–T algorithm is known as *decimation-in-time*. A similar technique divides the sequence for x_i directly into two interleaved sequences corresponding to the first $N/2$ samples and the last $N/2$ samples for x_i, i.e.

$$x_i = y_i'$$
$$x_{i+N/2} = z_i' \tag{4.33}$$

$(i = 0, 1, 2, \ldots , (N/2) - 1)$. This is known as *decimation-in-frequency*, and the transformed sequence X_n becomes

$$X_n = \sum_{i=0}^{(N/2)-1} (y_i' + z_i' W^{Nn/2}) W^{in}. \qquad (4.34)$$

The even and odd-numbered samples of X_n are calculated separately as

$$Y_n' = \sum_{i=0}^{(N/2)-1} (y_i' + z_i') W^{2in}. \qquad (4.35)$$

Replacing n by $2n + 1$ for n odd, we get

$$Z_n' = \sum_{i=0}^{(N/2)-1} (y_i' - z_i') W^i W^{2in} \qquad (4.36)$$

since $W^{Nn/2} = (-1)^n$.

Thus X_n is computed in terms of two DFTs of length $N/2$, described by eqns (4.35) and (4.36). As with the decimation-in-time algorithm, the same procedure is used recursively to compute the transformation in $\log_2 N$ stages. Note that the calculations for eqns (4.35) and (4.36) are similar except that eqn (4.36) requires a pre-multiplication by W^i of the input sequence. This does not affect the total number of arithmetic operations required for the complete transform, which remains the same as in the decimation-in-time case.

4.2.2 Matrices and the signal flow diagram

Orthogonal transformation can be considered in terms of matrix multiplications (eqn (3.42)) and for Fourier transformation can be expressed as

$$X_n = W^{in} x_i \qquad (4.37)$$

where W^{in} is a complex matrix having $N \times N$ terms. The key to fast matrix transformation lies in the manipulation of the matrix W^{in} to reduce the number of multiplications required. This is achieved by matrix factorization, considered in general terms in Section 3.3.5. The following simplified description of the process owes much to the procedures originally described by Theilheimer [22], and the author is grateful for permission to include these here.

A matrix for W^{in} is shown for $N = 8$ in Fig. 4.1. This may be simplified using eqn (4.30) and the general expression,

$$W^m = W^{m \bmod(N)} \qquad (4.38)$$

where $m \bmod(N)$ is the remainder when m is divided by N, i.e.

$$m \bmod(N) = m - |m/N| N \qquad (4.39)$$

$$
\begin{bmatrix} X_0 \\ X_1 \\ X_2 \\ X_3 \\ X_4 \\ X_5 \\ X_6 \\ X_7 \end{bmatrix}
\begin{bmatrix}
1 & 1 & 1 & 1 & 1 & 1 & 1 & 1 \\
1 & W & W^2 & W^3 & -1 & -W & -W^2 & -W^3 \\
1 & W^2 & -1 & -W^2 & 1 & W^2 & -1 & -W^2 \\
1 & W^3 & -W & W & -1 & -W^3 & W^2 & -W \\
1 & -1 & 1 & -1 & 1 & -1 & 1 & -1 \\
1 & -W & W & -W^3 & -1 & W & -W^2 & W^3 \\
1 & -W^2 & -1 & W^2 & 1 & -W^2 & -1 & W^2 \\
1 & -W^3 & -W^2 & -W & -1 & W^3 & W^2 & W
\end{bmatrix}
\begin{bmatrix} x_0 \\ x_1 \\ x_2 \\ x_3 \\ x_4 \\ x_5 \\ x_6 \\ x_7 \end{bmatrix}
$$

Fig. 4.3 Simplified DFT matrix.

and $|\,.\,|$ indicates an integer value (see Section 3.2.1). Hence for $N = 8$ we need not consider powers of W greater than three, so that the matrix shown in Fig. 4.1 reduces to Fig. 4.3 and we can write

$$X_n = F_8 x_i \qquad (4.40)$$

$(i, n = 0, 1, 2, \ldots, 7)$ and F_8 is an 8×8 DFT matrix. Using the procedure outlined in Section 3.3.4 this may be factorized into the form

$$X_n = [B_1 . C_1 . C_2 . P . P_1] x_i \qquad (4.41)$$

where B_1, C_1 and C_2 are sparse matrix factors of F_8, and P, P_1 are the permutation matrices.

P, P_1 are also sparse matrices and may be combined into a single matrix arranged to reorder (shuffle) either the input or output transformed values as a separate calculation operation.

Let $N = N_1 N_2 = 2 \times 4$. Then partitioning the top N_1 rows of F_8 into N_1 sections of N_2 coefficients for each section, i.e.

$$
\begin{array}{cccc|cccc}
1 & 1 & 1 & 1 & 1 & 1 & 1 & 1 \\
\hline
1 & W & W^2 & W^3 & -1 & -W & -W^2 & -W^3
\end{array}
\qquad (4.42)
$$

enables the first matrix, B_1, to be evolved from diagonal submatrices of value $N_2 \times N_2$ constructed from these partitioned sections

$$
B_1 = \left[
\begin{array}{cccc|cccc}
1 & 0 & 0 & 0 & 1 & 0 & 0 & 0 \\
0 & 1 & 0 & 0 & 0 & 1 & 0 & 0 \\
0 & 0 & 1 & 0 & 0 & 0 & 1 & 0 \\
0 & 0 & 0 & 1 & 0 & 0 & 0 & 1 \\
\hline
1 & 0 & 0 & 0 & -1 & 0 & 0 & 0 \\
0 & W & 0 & 0 & 0 & -W & 0 & 0 \\
0 & 0 & W^2 & 0 & 0 & 0 & -W^2 & 0 \\
0 & 0 & 0 & W^3 & 0 & 0 & 0 & W^3
\end{array}
\right].
$$

(4.43)

Matrix B_2 takes the form shown in eqn (3.26) as a right diagonal matrix having a series of submatrices of value $N_2 \times N_2$ situated along the diagonal. The elements for these submatrices are given as $W^{N.ab}$, where $a, b = 0, 1, 2, 3$.

Hence the indices for W will be (for $N_1 = 2$)

a	0	1	2	3
b				
0	0	0	0	0
1	0	2	4	6
2	0	4	8	12
3	0	6	12	18

so that using eqn (4.38) we can write for B_2

$$B_2 = \begin{bmatrix} \begin{array}{cccc|cccc} 1 & 1 & 1 & 1 & & & & \\ 1 & W^2 & -1 & -W^2 & & & 0 & \\ 1 & -1 & 1 & -1 & & & & \\ 1 & -W^2 & -1 & W^2 & & & & \\ \hline & & & & 1 & 1 & 1 & 1 \\ & & 0 & & 1 & W^2 & -1 & -W^2 \\ & & & & 1 & -1 & 1 & -1 \\ & & & & 1 & -W^2 & -1 & W^2 \end{array} \end{bmatrix}$$

(4.44)

where 0 indicates zero elements in all the submatrix element positions.

The reshuffling matrix P can be found from the elements of the first N_2 rows in matrix, B_1. The elements in these rows are taken row by row, giving each element in P the same column index as in B_1 but with the row index increased by a factor N_1 (apart from the first), i.e.

$$P = \begin{bmatrix} 1 & 0 & 0 & 0 & 0 & 0 & 0 & 0 \\ 0 & 0 & 0 & 0 & 1 & 0 & 0 & 0 \\ 0 & 1 & 0 & 0 & 0 & 0 & 0 & 0 \\ 0 & 0 & 0 & 0 & 0 & 1 & 0 & 0 \\ 0 & 0 & 1 & 0 & 0 & 0 & 0 & 0 \\ 0 & 0 & 0 & 0 & 0 & 0 & 1 & 0 \\ 0 & 0 & 0 & 1 & 0 & 0 & 0 & 0 \\ 0 & 0 & 0 & 0 & 0 & 0 & 0 & 1 \end{bmatrix}.$$ (4.45)

The submatrix B_2 is itself capable of factorizing into two matrices C_1 and C_2. Proceeding in the same way as for B_1 we can derive C_1 from the top two rows of the B_2 quadrant as

$$
C_1 = \left[
\begin{array}{cccc|cccc}
1 & 0 & 1 & 0 & & & & \\
0 & 1 & 0 & 1 & & & 0 & \\
1 & 0 & -1 & 0 & & & & \\
0 & W^2 & 0 & -W^2 & & & & \\
\hline
 & & & & 1 & 0 & 1 & 0 \\
 & & 0 & & 0 & 1 & 0 & 1 \\
 & & & & 1 & 0 & -1 & 0 \\
 & & & & 0 & W^2 & 0 & -W^2
\end{array}
\right]
$$

copyright © 1969 IEEE   (4.46)

Matrix C_2 will be a right diagonal matrix having a series of submatrices of value $N_2/2 \times N_2/2$ along the diagonal. The elements for these are W^{n2ab}, where a, $b = 0$, 1 and the indices for W are

a	0	1
b		
0	0	0
1	0	4

and

$$
C_2 = \left[
\begin{array}{cccccccc}
1 & 1 & & & & & & \\
1 & -1 & & & & 0 & & \\
 & & 1 & 1 & & & & \\
 & & 1 & -1 & & & & \\
 & & & & 1 & 1 & & \\
 & 0 & & & 1 & -1 & & \\
 & & & & & & 1 & 1 \\
 & & & & & & 1 & -1
\end{array}
\right]
$$

(4.47)

with a permutation matrix,

$$
P_1 = \left[
\begin{array}{cccc|cccc}
1 & 0 & 0 & 0 & & & & \\
0 & 0 & 1 & 0 & & & 0 & \\
0 & 1 & 0 & 0 & & & & \\
0 & 0 & 0 & 1 & & & & \\
\hline
& & & & 1 & 0 & 0 & 0 \\
& & 0 & & 0 & 0 & 1 & 0 \\
& & & & 0 & 1 & 0 & 0 \\
& & & & 0 & 0 & 0 & 1
\end{array}
\right]
\tag{4.48}
$$

The product of the sparse matrices, $B_1 \cdot C_1 \cdot C_2$ multiplied by the input vector x_i will give a transformed value X'_n in a bit-reversed order. The complete matrix multiplications for $N = 8$ are summarized in Fig. 4.4. The bit-reversal routine employing translation matrices, $P \cdot P_1$ is considered as a separate operation since non-matrix algorithms for bit-reversal can be easier to include in the computer program. Therefore the matrix equation, $X'_n = B_1 \cdot C_1 \cdot C_2 \cdot x_i$

will be taken as a practical basis for algorithmic development.

$$
\begin{bmatrix} X_0 \\ X_4 \\ X_2 \\ X_6 \\ X_1 \\ X_5 \\ X_3 \\ X_7 \end{bmatrix}
=
\begin{bmatrix}
1 & 0 & 0 & 0 & 1 & 0 & 0 & 0 \\
0 & 1 & 0 & 0 & 0 & 1 & 0 & 0 \\
0 & 0 & 1 & 0 & 0 & 0 & 1 & 0 \\
0 & 0 & 0 & 1 & 0 & 0 & 0 & 1 \\
1 & 0 & 0 & 0 & -1 & 0 & 0 & 0 \\
0 & W & 0 & 0 & 0 & -W & 0 & 0 \\
0 & 0 & W^2 & 0 & 0 & 0 & -W^2 & 0 \\
0 & 0 & 0 & W^3 & 0 & 0 & 0 & -W^3
\end{bmatrix}
\bullet
\begin{bmatrix}
1 & 0 & 1 & 0 & 0 & 0 & 0 & 0 \\
0 & 1 & 0 & 1 & 0 & 0 & 0 & 0 \\
1 & 0 & -1 & 0 & 0 & 0 & 0 & 0 \\
0 & W^2 & 0 & -W^2 & 0 & 0 & 0 & 0 \\
0 & 0 & 0 & 0 & 1 & 0 & 1 & 0 \\
0 & 0 & 0 & 0 & 0 & 1 & 0 & 1 \\
0 & 0 & 0 & 0 & 1 & 0 & -1 & 0 \\
0 & 0 & 0 & 0 & 0 & W^2 & 0 & -W^2
\end{bmatrix}
\bullet
\begin{bmatrix}
1 & 1 & 0 & 0 & 0 & 0 & 0 & 0 \\
1 & -1 & 0 & 0 & 0 & 0 & 0 & 0 \\
0 & 0 & 1 & 1 & 0 & 0 & 0 & 0 \\
0 & 0 & 1 & -1 & 0 & 0 & 0 & 0 \\
0 & 0 & 0 & 0 & 1 & 1 & 0 & 0 \\
0 & 0 & 0 & 0 & 1 & -1 & 0 & 0 \\
0 & 0 & 0 & 0 & 0 & 0 & 1 & 1 \\
0 & 0 & 0 & 0 & 0 & 0 & 1 & -1
\end{bmatrix}
\bullet
\begin{bmatrix} x_0 \\ x_1 \\ x_2 \\ x_3 \\ x_4 \\ x_5 \\ x_6 \\ x_7 \end{bmatrix}
$$

$$\underbrace{}_{B_1} \qquad\qquad \underbrace{}_{C_1} \qquad\qquad \underbrace{}_{C_2}$$

Fig. 4.4 Factorizing the DFT matrix.

The matrices, B_1, C_1, and C_2 are calculated in turn with the vector results in each stage used as input to the next matrix multiplication stage. For $N = 2^p$ then p matrix-vector product stages are necessary. A simple procedure exists that we can use to translate from this form of matrix evaluation to a complete algorithm, known as a *signal flow diagram*.

4.2.3   The signal flow diagram

The signal flow diagram [23] consists of a series of nodes, each representing a variable as the sum of a pair of variables originating from

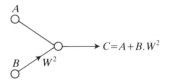

Fig. 4.5 Principle of the signal flow diagram.

the left of the diagram and with the nodes connected together by means of straight lines. These additions may be weighted by a number or symbol appearing at the side of an arrow on the line proceeding from the variable to be weighted (Fig. 4.5).

The procedure to convert a matrix representation into a signal flow diagram is described with reference to the first sparse matrix B_1, shown in eqn (4.43). This is reproduced in Fig. 4.6(a), where the columns have been labelled as the input series $x_0, x_1, \ldots, x_7$, and the rows are the intermediate output series $a_0, a_1, \ldots, a_7$. The single-stage flow diagram corresponding to this matrix is shown in Fig. 4.6(b). To select the pair of input terms to be combined to form a single output series it is necessary to scan the columns of the matrix to identify a pair of values other than zero along the output row. For example, row a_0 contains entries $+1$ and $+1$ for columns x_0 and x_4, enabling the appropriate lines connecting x_0 and x_4 to a_0 to be drawn on the flow diagram. Where the scan reveals a constant other than 1, e.g. 1 for x_0 and $-W$ for x_4 along row a_4, then the weighting factor $-W$ is included against the $x_4 - a_4$ line. The complete flow diagram for all three stages of the algorithm ($N = 2^3$) is given in Fig. 4.7.

The signal flow diagram gives a concise method for representing the computations required in the factorized matrix formulation of the FFT and other fast transformations. Using this graphical presentation allows the matrix factorization process for large values of N to be visualized. Inspection of a flow diagram for $N = 16$ or 32 permits the general

(a)

	Input series								
x_0	x_1	x_2	x_3	x_4	x_5	x_6	x_7		
1	0	0	0	1	0	0	0	a_0	
0	1	0	0	0	1	0	0	a_1	
0	0	1	0	0	0	1	0	a_2	
0	0	0	1	0	0	0	1	a_3	Output
1	0	0	0	-1	0	0	0	a_4	series
0	W	0	0	0	$-W$	0	0	a_5	
0	0	W^2	0	0	0	$-W^2$	0	a_6	
0	0	0	W^3	0	0	0	$-W^3$	a_7	

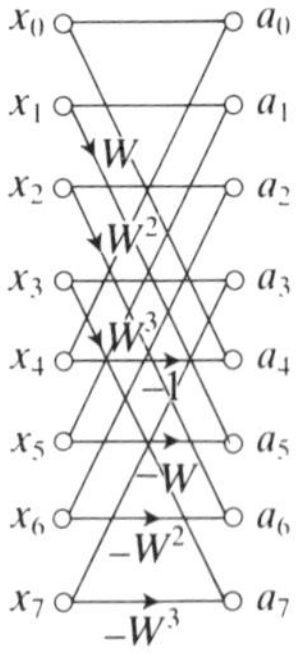

Fig. 4.6 Conversion from a matrix to a signal flow diagram.

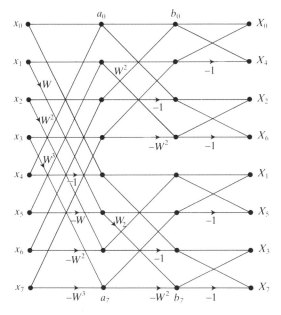

Fig. 4.7 Flow diagram for a S–T FFT algorithm (bit-reversed output).

combinational requirements to be understood, leading to the development of a computer flow chart.

Figure 4.7 shows the flow diagram of the S–T algorithm for $N = 8$. The output is in bit-reversed order. A decimation-in-frequency algorithm giving an output in sequential order but input in bit-reversed order is shown in Fig. 4.8. The C–T decimation-in-time algorithm for output in bit-reversed order, considered in some detail earlier in algebraic terms, is shown in Fig. 4.9. Its alternative, with a bit-reversed input, is given in Fig. 4.10. Note that Figs 4.7 and 4.10 provide the powers of W in the correct order for computation. This can remove the need for storage tables for W values in the computer program.

A common feature of these diagrams is that they are built up from a series of simpler geometries, two of which are shown in Fig. 4.11. These have become known by their shape as *butterflies* and represent the basic operation of producing sums and differences from pairs of terms. Figure 4.11(a) illustrates a butterfly carrying out the operation $A + B \cdot W^n$, such as we find in the decimation-in-time algorithms (eqns 4.28 and 4.31) and Fig. 4.11(b) illustrates the operation $(A + B)W^n$ found in the decimation-in-frequency algorithms. These basic operations imply that once the summations $(A + B)$ have been determined from A and B then the initial data A and B are no longer required. Algorithms composed entirely from butterflies such as those shown in Figs 4.8–4.11 are known as *in-place*

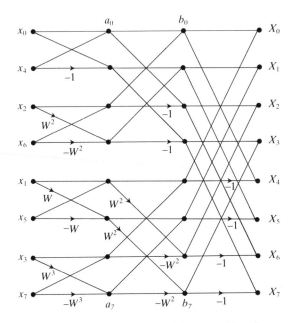

Fig. 4.8 Flow diagram for a S–T FFT algorithm (bit-reversed input).

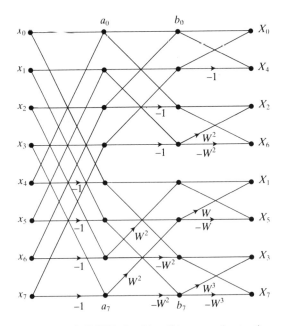

Fig. 4.9 Flow diagram for a C–T FFT algorithm (bit-reversed output).

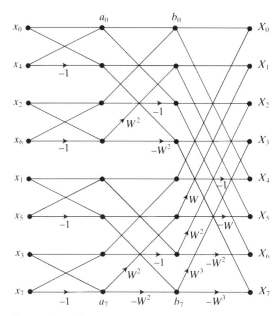

Fig. 4.10 Flow diagram for a C–T FFT algorithm (bit-reversed input).

algorithms. Thus it is not necessary to allocate separate computer memory storage for the intermediate arrays of values, a_0, b_0 etc. Storage is only required to accommodate the initial number of data points and a similar area for the calculation product values (since each calculated value is needed twice). The first set of intermediate transforms can be calculated sequentially and placed in the locations previously occupied by

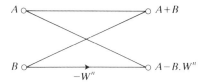

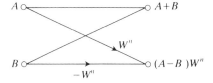

Fig. 4.11 Butterfly diagrams.

the first set of values, which are no longer needed. The second set can then be calculated and returned to the locations occupied by the first set and so on until the transformation is completed. Algorithms exist in which both input and transformed output data can be in the correct natural order. However in this case the computations can no longer be carried out in-place and at least double the array storage is necessary [24].

4.2.4 Twiddle factor algorithms

Significant additional reductions in computing time can be achieved if the symmetries in the sine and cosine functions implied in the twiddle factors W^m are taken into account. Thus for the radix-2 C–T algorithm for $N = 8$ shown in Fig. 4.9 we have already noted that the twiddle factors in the first stage a_0 are given by $W^{nN/2} = (-1)^n$ so that all multiplications in this stage are trivial. Similarly we find that the multiplications in the second stage are also trivial since the twiddle factors are defined by $W^{nN/4} = (-j)^n$. Non-trivial multiplications in subsequent stages can be reduced by considering the twiddle factors as products of these simple forms and an integer, reducing the number of the multiplications to $+1$ or $+j$. It can be shown that under these conditions the number of non-trivial multiplications required in a radix-2 algorithm of N values will be,

$$N/2(-3 + \log_2 N) + 2 \tag{4.50}$$

which is, for example, about half that required for $N = 1024$ when eqn (4.27) is calculated directly.

Similar advantages are obtained for radix-4, radix-8 and radix-16. While performing as many high radix computations as possible will reduce the number of multiplications required the algorithm for the higher radices becomes more involved. Radix-4 and Radix-8 appear to be the most efficient without increasing the complexity too much [25].

4.2.5 Inverse transformation

From the definitions of the DFT and the IDFT given in Chapter 2 we can easily conclude that the FFT algorithm can be used also for calculation of the inverse discrete Fourier transform. This is in fact the case. The IDFT is defined as,

$$x_i = \sum_{n=0}^{N-1} X_n W^{-in} \tag{4.51}$$

$(n = 0, 1, 2, \ldots, N - 1)$. If we take the complex conjugate of eqn (4.51) and multiply by N we obtain,

$$x_i^* = \sum_{n=0}^{N-1} X_n^* W^{in} \tag{4.52}$$

The right-hand side of eqn (4.52) will be recognized as the DFT of X_n^* and may be computed using any of the algorithms described previously without any change. We need to carry out a complex conjugate process on the output as well as dividing N to give

$$x_i = 1/N \left[\sum_{n^*0}^{N-1} X_n^* W^{in} \right]^*. \tag{4.53}$$

4.2.6 Transformation of real data

The basic FFT algorithms described above have been developed for complex data series. When applied to signal analysis real data will be available and hence transforms of a real data series will be needed. We can, of course, set the imaginary parts to zero and insert the relevant data samples in the real part of the synthesized complex sequence. However, this will be wasteful both of computer time and storage space. A more efficient procedure makes use of the conjugate symmetry of a real variable. If a data series x_i is real, then

$$x_i = x_i^*$$

and

$$X_n = X_{-n}^* = X_{N-n}^* \tag{4.54}$$

This latter relationship follows from the periodic nature of the exponent W in eqn (4.2)

$$W^{in} = W^{(n+N)i} = W^{(i+N)n}. \tag{4.55}$$

Hence,

$$x_i = x_{kN+i}$$

$$X_n = X_{kN+n}$$

$(k = 0, 1, 2, \ldots)$. Therefore

$$x_{-i} = x_{N-i}$$

$$X_{-n} = X_{N-n}$$

$$X_{-n}^* = X_{N-n}^* \tag{4.56}$$

This means that if x_i is real then we only need *half* the data points in order to specify the frequency spectrum.

If we consider the real data series to be separated into two interleaved series, as given in eqn (4.24) then defining y_i as the real part, and z_i as the imaginary part of a complex function x_i we can write $x_i = (y_i + jz_i)$. The DFT of this will give (neglecting scaling by $1/N$),

$$X_n = \sum_{i=0}^{(N/2)-1} (y_i + jz_i)W^{2in}$$

$$= \sum_{i=0}^{(N/2)-1} y_i W^{2in} + j \sum_{i=0}^{(N/2)-1} z_i W^{2in} \tag{4.57}$$

which we can represent as,

$$X_n = Y_n + jZ_n. \tag{4.58}$$

But from eqn (4.28)

$$X_n = Y_n + W^n Z_n \tag{4.59}$$

$(n = 0, 1, 2, \ldots, (N/2) - 1)$. Each of these transforms will contain $N/2$ points, and since the original series y_i and z_i were real, then from eqn (4.54) only $N/4$ points will be required to describe them. Hence,

$$Y_{(N/4)-m} = y^*$$

and

$$Z_{(N/4)-m} = Z^*_{(N/4)+m} \tag{4.60}$$

$(m = 0, 1, 2, \ldots, (N/4) - 1)$. Also,

$$X_n^* = Y_n - jZ_n \tag{4.61}$$

so that by adding and subtracting eqns (4.56) and (4.61) we can obtain,

$$X_n + X_n^* = 2Y_{(N/4)-m}$$

and

$$X_n - X_n^* = 2jZ_{(N/4)-m}. \tag{4.62}$$

This enables Y_n and Z_n to be obtained from our original transform in X_n and substituted in eqn (4.28) to obtain X_n for the first $N/4$ points.

The second $N/4$ points are obtained from

$$X_{(N/2)-m} = Y_n - W^n Z_n. \tag{4.63}$$

A similar procedure can be carried out to evaluate simultaneously the Fourier transforms of two equal-length sets of real data. Thus if we let the two series be x_1 with its Fourier transform X_1, and x_2 with its Fourier transform X_2, then substituting these arrays to form the real and imaginary parts of a complex series, x_i (note that an interleaved series is not involved here),

$$X_n = X_1 + jX_2. \tag{4.64}$$

From eqn (4.54) we can replace n by $N - n$ so that by taking the complex conjugate of both sides we obtain

$$X_{N-n}^* = X_1 - jX_2 \tag{4.65}$$

giving

$$X_1 = 1/2[X_n + X_{N-n}^*]$$
$$X_2 = 1/2[X_n - X_{N-n}^*]. \tag{4.66}$$

Since X_1 and X_2 are real, the symmetry property of eqn (4.54) applies and only half of each array need be computed and stored. Thus only the same amount of storage will be required to evaluate X_1 and X_2 taken together as would be needed to obtain a single complex transform X_n containing half the number of points. Computation time will be only slightly increased by the necessity to form sums and differences using eqn (4.66).

4.2.7 Transforms for radix-4, radix-8, and mixed radix

A reduction in numbers of multiplications and additions can be realized at the expense of some program complication by using other powers of two for the radix. Successful implementations of radix-4 and radix-8 algorithms are in wide use [10, 11]. The increase in complexity for higher radices (e.g. radix-16) has not been found to be justified in terms of computation speed [11].

Consider an FFT where $N = 2^p$ and p is even. The N-point series for x_i is considered separated into four interleaved series of $N/4$ values

$$x_{4i}, x_{4i+1}, x_{4i+2}, \quad \text{and} \quad x_{4i+3} \tag{4.67}$$

where $i = 0, \ldots, (N/4) - 1$. The transformation becomes

$$X_n = \sum_{f=0}^{3} W^{fn} \sum_{i=0}^{(N/4)-1} x_{4i+f} W^{4in} \tag{4.68}$$

but since $W^{N/4} = -j$ then

$$X_{n+N/4} = \sum_{f=0}^{3} -j^f \cdot W^{fn} \sum_{i=0}^{(N/4)-1} x_{4i+f} W^{4in} \tag{4.69}$$

$$X_{n+N/2} = \sum_{f=0}^{3} -1^f W^{fn} \sum_{i=0}^{(N/4)-1} x_{4i+f} W^{4in} \tag{4.70}$$

$$X_{n+3N/4} = \sum_{f=0}^{3} j^f W^{fn} \sum_{i=0}^{(N/4)-1} x_{4i+f} W^{4in}. \tag{4.71}$$

Thus the radix-4 decimation-in-time algorithm for the DFT may be shown to reduce to four DFTs each of length $N/4$. The same procedure may be applied recursively in $p/2$ stages, each stage reducing the dimensions of the DFT by a factor of 4. As defined above this radix-4 algorithm would result in a number of complex operations actually higher than a radix-2 algorithm. However by making use of the 'twiddle-factor' reductions (see Section 4.2.4) and noting that for some of the calculation stages the twiddle factors reduce to trivial multiplications by ± 1 and $\pm j$ considerable reduction in computation time becomes possible [13]. A comparison of operations required in computing a C–T algorithm for radix-2, radix-4, radix-8, and radix-16 is given in Table 4.1 (after Bergland).

When N is not a power of a single radix it may be advantageous to vary the radix for different stages in the algorithm. The procedures that result from this approach are known as *mixed-radix algorithms*. It is possible to achieve somewhere near optimum performance for the C–T or S–T algorithms in this way, having regard to the number of arithmetic operations required, but the implementation will be found to increase in complexity compared with the simpler radix-2 algorithms [11, 12].

4.2.8 Bit-reversal

'In-place' algorithms of the type we have been discussing require that either the input or the output data sequences must be reordered. This reordering is equivalent to the bit-reversal of the sample index when this is expressed as a binary number. This was illustrated for $N = 8$ in Table 3.2, where the natural-ordered indices are as shown at the left and the bit-reversed indices resulting from binary representation and its inversion are shown on the right.

This reordering or *shuffling* of the indices can also be carried out 'in-place', i.e. pairs of numbers can be interchanged using only one temporary storage location. A flow diagram for $N = 8$ is shown in Fig. 4.12. A number of bit-reversal algorithms have been suggested for this operation. An easily implemented bit-reversing counter is given by Rader

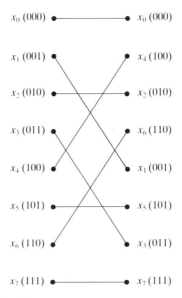

Fig. 4.12 Bit-inversion algorithm.

[26]. Other methods carry out the reversal in steps where each step corresponding to the exchange of groups of symmetric bits [27]. This latter is particularly efficient for large arrays because it minimizes the number of transfers taking place.

4.3 Reduced multiplication algorithms

The algorithms described by Cooley, Tukey, Sande, and others were developed in the 1960s using matrix factorization, matrix-analysis or equivalent series manipulation. These methods, leading to radix-2 fast transformations, have been discussed in earlier parts of this chapter. Two difficulties arise with these implementations. First, the constraint limiting the transformation of series expressed as a power of two can be restrictive. Second, the large number of complex multiplications required limit the speed of operation for the fast transformation. It may be noted here that these multiplications are associated with the twiddle factor W^n required where the algorithm is assembled from a number of common butterfly operations (eqns (4.28) and (4.31)). Since the number of real multiplications per point is almost the same for both real and complex input data, the FFT becomes very inefficient for calculating the FDT of real input data. The multiplication problem is significant, as we see when considering the multiplication of two N-bit numbers. This requires N additions per digit and for any reasonable accuracy; (say an 8-bit word) the product of two words is the result of adding eight binary numbers to give one real number product. Multiplication can therefore consume the majority of transform computer time, especially for long word lengths where high accuracy is needed.

For these reasons considerable activity in fast transform development has recently been related to the search for algorithms that require fewer multiplication operations. Early ideas led to the use of radices other than two and to mixed radix transforms. One of the most successful of these is Singleton's mixed radix algorithm [12], which is particularly effective when N is highly factorable by 2, 3, 4, or 5 and requires little additional memory relative to N.

The performance of the C–T-based algorithms, even with the flexibility obtained by radix adjustment, has now been overtaken by later developments in reduced multiplication and number-theoretic algorithms. Current hardware trends are likely to make these even more important in the future.

Reduced multiplication algorithms arose from work carried out by Good and Thomas [8, 28, 29] together with a revival of some early ideas in number theory being discussed by mathematicians [15]. The first applicable algorithm was described by Winograd in 1975 [30]. The two basic concepts of these and later algorithms are the reduction in multiplication operations required when N is restricted to the product of several small prime numbers, and the techniques used to make the FFT operation look like a circular convolution calculation. The former leads to a number of compact prime factor algorithms that in certain cases can eliminate multiplication altogether [31], and both techniques are used in the highly efficient Winograd and prime factor algorithms.

A further development in the late 1970s was the construction of fast algorithms for computing DFTs using polynomial transforms and other *number-theoretic transforms,* which include the Mersanne, Fermat, and Rader transforms [17, 32–34] following some very early and previously neglected work by Pollard [35].

These number theoretic transforms have been shown to be highly efficient for evaluating circular convolution but have not yet found wide use in signal processing operations due to their high overheads in program size and the 'book-keeping' routines required. The Mersanne transform does not possess a fast transform algorithm for example, and a large number of shift-additions are needed to construct the FDT. The Fermat number transform looks the most promising, particularly for hardware design, since its implementation requires only $N \log_2 N$ additions, subtractions and bit-shifts and no multiplications [36, 37]. It is however limited to use with relatively short data sequences.

In general both these sets of newer algorithms do not have the in-place calculation features of the C–T and S–T algorithms and a special kind of shuffling of the input or output data is required. The large amounts of accessing and storing of data, as well as loop control overheads and the complexity of these algorithms, put them at a disadvantage compared with the radix-2 transforms considered in Section 4.2. However for real-time operation or transformation of large amounts of data the savings in computational time can be considerable, and as noted earlier the newer algorithms are likely to find favour in hardware designs since they are more suitable for modular and parallel operation [38].

4.3.1 Prime factor algorithms

It will have been realized from earlier discussions that the choice of N for the single-dimensional data series has a profound effect on the kind of transformation routine that can be employed. If N is restricted to be a power of two then a simple but efficient factorization is possible, leading to the C–T and S–T FFT algorithms. A different restriction, in which N is made a prime number or the product of a number of relative prime numbers, can also be made to yield manipulative advantages.

The *prime factor FFT algorithm* (PFA) is a computation technique that allows one to compute a large DFT of size N by combining several small DFTs of sizes, $N_1, N_2, \ldots, N_p$, which are relative prime factors of N.

Consider first the simplest case in which $N = N_1 N_2$, with N_1 and N_2 being mutually prime factors. From eqn (4.2) we have for the DFT

$$X_n = \sum_{i=0}^{N-1} x_i W^{in} \qquad (4.72)$$

$(i, n = 0, 1, 2, \ldots, N-1)$. To express this one-dimensional DFT as a two-dimensional DFT of size $N_1 \times N_2$ we need to map from one- to two-dimensional indices just as we did earlier through eqns (4.3) and (4.4).

Expressing this in a more general form in terms of the mod. N_1 and mod. N_2 values (eqn (4.39)), we first relate the index i into a pair of indices, (i_1, i_2), i.e.

$$i_1 = r_1 i \bmod N_1$$

$$i_2 = r_2 i \bmod N_2 \qquad (4.73)$$

where $r_1 = N_2 \bmod N_1$ and $r_2 = N_1 \bmod N_2$. Second, we relate index n into a pair of indices (n_1, n_2), i.e.

$$n_1 = n \bmod N_1$$

$$n_2 = n \bmod N_2$$

The inverse mapping from two dimensions to one dimension is,

$$n = (s_1 n_1 + s_2 n_2) \bmod N$$

where $s_1 = 1 \bmod N_1$ and $s_2 = 0 \bmod N_1$

$$s_1 = 0 \bmod N_2 \quad \text{and} \quad s_2 = 1 \bmod N_2 \qquad (4.74)$$

and

$$i = (N_2 i_1 + N_1 i_2) \bmod N. \qquad (4.75)$$

Using these mappings with eqn (4.72) gives a two-dimensional representation for the DFT as

$$X_{n_1, n_2} = \sum_{i_1=0}^{N_1-1} \sum_{i_2=0}^{N_2-1} x_{i_1, i_2} W_{N2}^{i_2, n_2} W_{N1}^{i_1, n_1} \qquad (4.76)$$

where $W_{N1} = \exp(-j2\pi/N1)$, and $W_{N_2} = \exp(-j2\pi/N2)$ The two-dimensional transform of eqn (4.76) may be implemented by first calculating N_1 DFTs each of length N_2, i.e.

$$y_{i_1, n_2} = \sum_{i_2=0}^{N_2-1} x_{i_1, i_2} W^{i_2, n_2} \qquad (4.77)$$

and then calculating N_2 DFTs, each of length N_1, i.e.

$$X_{n_1,n_2} = \sum_{i_1=0}^{N_1-1} y_{i_1,n_2} W^{i_1,n_1}. \tag{4.78}$$

The calculations required to evaluate eqn (4.76) in this way can be shown to be fewer than those required for the original DFT (eqn (4.67)) [8].

If we compare eqns (4.77) and (4.78) with eqns (4.29) and (4.31), which result from the successive doubling method for an FFT, we note that the additional twiddle factor W^n is absent. Neither the calculation of Y_{i_1,n_2} or X_{n_1,n_2} require multiplication by W^n as with Z_n in eqns (4.29) and (4.31). This explains why the PFA is more efficient, since the twiddle factors involve exponents n_1, i_2 or n_2, i_1 and require considerable additional multiplication for the DFT computation.

The price to pay for this simplification, apart from the restriction in the prime number factorizing required for N, is that the order of the input and output sequences is shuffled and the calculations are not performed in-place. This shuffling is not a simple bit-reversal as described in Section 4.2.8 but something rather more complicated. Its implementation depends on the actual factorization chosen for N and is not necessarily the same for input and output sequences. To determine the mapping algorithm needed in a given case use is made of a *Chinese remainder theorem*. For a full description of this the reader is referred to McClennon and Rader [15] and Saucedo and Schiring [23].

An alternative PFA has been suggested by Burrus and Eschenbacher, which carries out in-place calculation and presents the output in natural sequential order [38]. This requires a modification for each of the factorized DFT algorithms used and the program construction employed is therefore made dependent on the size of N. To use the algorithm therefore it is necessary to add a fairly complicated set of control statements or to recompile the program for each new value chosen for N. A way of overcoming this latter difficulty, that removes the need for separate input and output arrays is given by Rothweiler [39].

The PFA example given above considers only two dimensions in N, namely N_1 and N_2. For higher dimensions the same approach is taken, with N factorized into relative prime numbers.

It is possible to show a similar factorization using powers of prime numbers, e.g. 2^2, 3^3 etc. The number of multiplications M and additions A required using a PFA algorithm are given by

$$M = N_1 M_2 + N_2 M_1$$
$$A = N_1 A_2 + N_2 A_1 \tag{4.79}$$

where $N = N_1 N_2$ and N_1, N_2 are relatively prime. $A_1 M_1$ is associated with the DFT of length N_1 and $A_2 M_2$ with N_2.

Extending eqn (4.79) recursively to cover the case of D factors we have

$$N = \prod_{k=1}^{D} N_k \tag{4.80}$$

where $N_1, N_2, \ldots, N_D$ are relatively prime, and

$$M = \sum_{k=1}^{D} NM_k/N_k \tag{4.81}$$

$$A = \sum_{k=1}^{D} NA_k/N_k. \tag{4.82}$$

Comparisons with radix-2 algorithms indicate a reduction in both M and A of the order of 50 per cent for the PFA where $N > 512$.

A PFA FFT which achieves high-speed operation not by reduction in the number of multiplications required but by reduction in calculation overheads is described by Chu and Burrus, and termed here the C–B PFA [40]. In this algorithm the DFT is first decomposed into multi-dimensional DFTs using a prime factor index. These small-N DFTs are then converted into convolutions as discussed in the next section (4.3.2), and finally the convolutions are calculated by distributed arithmetic using table look-up to access a set of pre-calculated constants.

It is this last stage that makes the C–B PFA different from the earlier PFAs [16, 34] and the WFTA [14]. Pairs of signal input values $a_i = x_i + x_{N-i}$ are multiplied by the pre-computed multiplier values (essentially a matrix of sine–cosine functions W_r held in a ROM look-up table). The products are output to a set of $N/2 - 1$ accumulators connected as a circular shift register. Consider the case where $N = 12$.

Initially the accumulators are set to x_0. Next a_1 times the first column for the transform matrix W_r is added to these five accumulators. The contents of all the accumulators are circularly shifted up and the same operation is carried out with a_5 times the second column of W_r. This continues until all input pairs have been dealt with, leaving the final convolved product summation in the output accumulators. Thus a parallel operation is carried out to evolve the small-N convolutions in which the use of a large memory for the ROM look-up table replaces the need for repetitive sequential calculations.

Using a Z80 microprocessor and software program control a speed improvement for the FFT of between 2 and 20 times is obtained compared with an efficient radix-2 algorithm. It has been pointed out by Chu and Burrus that since the distributed arithmetic method consists of a regular repetition of fetch and accumulate operations they can be carried out automatically by means of additional hardware to obtain an improvement on these performance figures. Details of this external hardware (an additional accumulator and control system) is given in [40].

This technique of factorizing N into relative prime or powers of prime numbers is also the basis for other reduced multiplication algorithms such as Winograd, where the DFT is constructed of any permissible order N through the sequential application of special generalized DFT algorithms for the following eight orders

$$N_k = 2, 3, 4, 5, 7, 8, 9, 16.$$

This limits the maximum size of N to $N = 5040$ (since 5, 7, 9, and 16 are the largest mutually prime numbers in this set). Since the size of a constituent transform is limited to these very small numbers it is possible to consider forms of algorithm to give the absolute minimum number of multiplications in each case. These 'tailored' algorithms are known as *short DFT algorithms* and are the building blocks for the PFA, Winograd and other reduced multiplication transforms.

4.3.2 Short DFT algorithms

A key idea for minimum multiplications for short DFT algorithms is described by Rader [34]. He observed that the computation of the DFT can be changed into circular convolution by rearranging the data matrix when N is a prime number. The minimum number of multiplications required for circular convolution in a given N is known through the work of Winograd [14], and several convolution algorithms that achieve this minimum are available [14, 16, 20].

We consider first the matrix representation of the DFT given by eqn (4.51), where N is a prime. This was shown earlier in Fig. 4.1 as a linear transformation of an N-dimensional data vector x_i into a vector X_n of frequency samples. Examination of the matrix for W will show that the complexity of an algorithm for calculating this matrix arises partly from the need to include the many W^0 terms in the first row and column of the matrix. These however only contribute to the mean level of the result and we can compute eqn (4.2) from

$$X_n = X_0 + \sum_{i=1}^{N-1} x_i W^{in} \qquad (4.83)$$

$(i, n = 1, 2, \ldots, N - 1)$, where X_0 represents a mean or average level for the discrete series x_i, i.e.

$$X_0 = \sum_{i=0}^{N-1} x_i \qquad (4.84)$$

so that the reduced transformation can be expressed as

$$X_n' = \sum_{i=1}^{N-1} x_i W^{in}. \qquad (4.85)$$

To see how eqn (4.85) may be converted to circular convolution, consider a matrix representation for $N = 5$ with exponents taken as modulo-5 (eqn (4.38))

$$\begin{bmatrix} X_1' \\ X_2' \\ X_3' \\ X_4' \end{bmatrix} = \begin{bmatrix} W^1 & W^2 & W^3 & W^4 \\ W^2 & W^4 & W^1 & W^3 \\ W^3 & W^1 & W^4 & W^2 \\ W^4 & W^3 & W^2 & W^1 \end{bmatrix} \cdot \begin{bmatrix} x_1 \\ x_2 \\ x_3 \\ x_4 \end{bmatrix}. \qquad (4.96)$$

As we saw in Chapter 3 it is permissible to interchange rows and columns in a matrix without changing its properties. Interchanging the last two columns of eqn (4.86) and then interchanging the last two rows gives,

$$\begin{bmatrix} X_1' \\ X_2' \\ X_4' \\ X_3' \end{bmatrix} = \begin{bmatrix} W^1 & W^2 & W^4 & W^3 \\ W^2 & W^4 & W^3 & W^1 \\ W^4 & W^3 & W^1 & W^2 \\ W^3 & W^1 & W^2 & W^4 \end{bmatrix} \cdot \begin{bmatrix} x_1 \\ x_2 \\ x_4 \\ x_3 \end{bmatrix}. \qquad (4.87)$$

This new orthogonal matrix is similar to a circular correlation matrix but not exactly equal (see Section 3.4.3). It is necessary to achieve equality to ensure that the multipliers are shifted to the right in moving from one row to the next one down. This is obtained by reversing the order of the second and fourth columns to give

$$\begin{bmatrix} X_1' \\ X_2' \\ X_4' \\ X_3' \end{bmatrix} = \begin{bmatrix} W^1 & W^3 & W^4 & W^2 \\ W^2 & W^1 & W^3 & W^4 \\ W^4 & W^2 & W^1 & W^3 \\ W^3 & W^4 & W^2 & W^1 \end{bmatrix} \cdot \begin{bmatrix} x_1 \\ x_3 \\ x_4 \\ x_2 \end{bmatrix}. \qquad (4.88)$$

A more rigorous derivation of the circulant matrix is given by Rader in terms of a rule involving the primitive root of N, where N is a prime number. The short DFT algorithms arising out of this method are called *Rader's algorithms*. It may also be shown that the DFT can be converted into a circular convolution if N is a prime power, i.e. $N = p^r$ for a prime $p = 2$ [14].

Once the prime or prime power DFT has been expressed as a convolution it is possible to employ fast cyclic convolution techniques to perform the calculations. These have been shown to reduce the number of complex multiplications required, from N^2 in the case of eqn (4.51), to approximately N.

Convolution algorithms that achieve this performance are only known for short lengths of N, and a set of these has been given elsewhere by Winograd for $N = 2, 3, 4, 5, 7, 9$, and 16. [14]. Two of these are reproduced below for $N_1 = 3$ and $N_2 = 4$, with the number of arithmetic operations as shown. The equivalent signal flow diagrams for the algorithms are given in Figs 4.13 and 4.14, respectively.

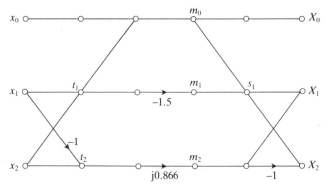

Fig. 4.13 Flow diagram for a prime number FFT with $N_1 = 3$.

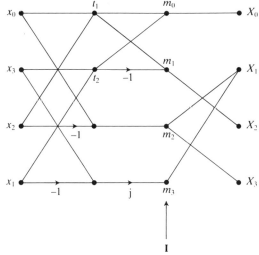

Fig. 4.14 Flow diagram for a prime number FFT with $N_1 = 4$.

Three-point DFT   (three multiplications and six additions)

$$t_1 = x_1 + x_2 \qquad\qquad t_2 = x_2 - x_1$$
$$m_0 = 1 \cdot (x_0 + t_1) \qquad m_1 = -1 \cdot 5t_1$$
$$m_2 = j0.866t_2 \qquad\qquad s_1 = m_0 + m_1$$
$$X_0 = m_0 \qquad\qquad\qquad X_1 = s_1 + m_2$$
$$X_2 = s_1 - m_2. \qquad\qquad\qquad\qquad\qquad (4.89)$$

Four-point DFT   (four multiplications and eight additions)

$$t_1 = x_0 + x_2 \qquad\qquad t_2 = x_1 + x_3$$
$$m_0 = 1(t_1 + t_2) \qquad\quad m_1 = 1(t_1 - t_2)$$
$$m_2 = 1(x_0 - x_2) \qquad\quad m_3 = j(x_3 - x_1)$$
$$X_0 = m_0 \qquad\qquad\qquad X_1 = m_2 + m_3$$
$$X_2 = m_1 \qquad\qquad\qquad X_3 = m_2 - m_3. \qquad (4.90)$$

4.3.3 The Winograd Fourier transform

This was the earliest and is probably still the most important of the reduced multiplication Fourier transform algorithms. Winograd contributed two developments which were new, or not properly appreciated at the time. The first was to take a data sequence

$$N = \prod_{i=1}^{k} N_i$$

(where the N_i are relatively prime) and to turn this into N/N_i sequences of short circular convolutions of length $N - (i)$ (for $i = 1, \ldots, k$) using an index mapping described by Good [8]. The second was to combine these short algorithms through a nesting structure to produce the required transformation of N values. It is this latter process that has been termed the Winograd Fourier transform algorithm (WFTA). This algorithm reduces the number of multiplications to about 20 per cent of those needed in an equivalent C–T FFT while leaving the number of additions at about the same level.

The WFTA is applicable to any N satisfying

$$N = \prod_{i=1}^{D} N_k \qquad (4.91)$$

in which N_k are relative prime factors taken from the list $N_k = 2, 3, 4, 5, 7, 8, 9, 16$. To assess the actual gain in calculation time we need to know the ratio

$$\frac{\text{Time to execute one real multiplication}}{\text{Time to execute one real addition}} \qquad (4.92)$$

which will be different for different systems and is particularly large for microprocessors and software multipliers where the savings in numbers of multiplication operations is particularly effective. This indicates comparative savings for the arithmetic operations only, and the structural complexity of Winograd's algorithm will tend to slow down its software implementation. The main disadvantage of the algorithm apart from its more complex software lies in its memory requirements. Storage array in real words is dependent on the size of N. This actually varies between about $3N$ to $6N$ and includes an allocation of about one third of this for the storage of the pre-computed constants.

The WFTA consists of five distinct procedures;

1. The permutation of the signal data sequence into k mutually prime integers.

2. A 'pre-weave' operation to carry out just the additions and subtractions required for each of the k small-N transforms. This will also include the necessary data transfers and loop overheads.

3. A point-by-point multiplication of the data array with an array of real constants derived from the multipliers of the small-N transforms.

4. A 'post-weave' operation, one for each k value, to reduce the data array back to its original size.

5. A shuffling or 'mapping' of the k-dimensional array into a one-dimensional DFT data array.

To derive the WFTA from the small-N DFTs these need first to be converted to a factored matrix representation. This is equivalent to matrix factorization to derive the FFT. With $N = 2^p$, then p factored matrices are required for the radix-2 FFT. These are programmed in terms of in-place computation of butterflies rather than storing and manipulating each of the sparse matrices. With the WFTA the factored matrices representing the small-N DFTs, which are also sparse matrices having many zero terms, are not stored in memory. Instead the equations that carry out the matrix operations are stored and operated upon directly. These equations do not have the symmetrical form of the radix-2 FFTs and therefore require more program storage.

The basic operation of the small-N DFTs is to separate out the addition of input terms, multiplication of the resulting sums, and the addition of the results. These form the three quite distinct procedures (2), (3), and (4) listed above. Thus if S represents the matrix form of a small DFT we can represent its computation by

$$S = xXM \qquad (4.93)$$

where x carries out input additions, M carries out output additions, and X carries out all multiplications

What this does is to nest all the multiplications inside the input and output additions.

These various matrix operations are shown diagrammatically in Fig. 4.15. Once the input order shuffling has taken place the input additions on the rows of the input values are carried out followed by the column additions. Multiplication is carried out through a diagonal matrix which is

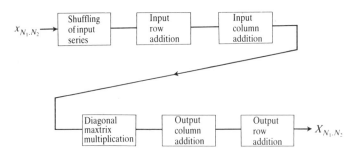

Fig. 4.15 Nesting operations for matrix multiplication.

effectively a scalar operation. Finally row and column addition are performed on the output values to produce the transform coefficients.

As stated above the Winograd fast algorithm is developed from small-N DFTs by taking the product of a number of small-N matrices nested in such a way as to group all the input additions of the large N ($N = N_1, N_2, \ldots, N_D$) separately from the transform multiplication and output additions. This is achieved through their Kronecker product expansions (see Section 3.3.4) using the relationship

$$(AB) \otimes (CD) = (A \otimes C) \,.\, (B \otimes D) \tag{4.94}$$

where A, B, C, and D are rectangular matrices. Thus for $N = N_k, \ldots, N_2 N_1$

$$D = (X_k M_k x_k) \otimes \ldots \otimes (X_2 M_2 x_2) \otimes (X_1 M_1 x_1). \tag{4.95}$$

Using eqn (3.46) we can regroup this as

$$D = (X_k \underbrace{\otimes \ldots \otimes X_2 \otimes X_1}_{\text{output additions}}) \,.\, (M_k \underbrace{\otimes \ldots \otimes M_2 \otimes M_1}_{\text{multiplications}})$$

$$\,.\, (x_k \underbrace{\otimes \ldots \otimes x_2 \otimes x_1}_{\text{input additions}}). \tag{4.96}$$

The Winograd algorithm separates the computation of a DFT of size $N = N_k, \ldots, N_2 N_1$ into the evaluation of small DFTs of length $N_1, N_2, \ldots, N_k$ in a manner that is fundamentally different from the PFA discussed earlier, although both start from the idea of fast small-N prime factor algorithms. A complete flow diagram for the WFTA is shown in Fig. 4.16 for $N = N_1 \,.\, N_2 = 3 \times 4$. This shows clearly the effect of nesting on the design of the complete algorithm with the four repetitions of the $N_1 = 3$ flow diagram inserted at point I in the $N_1 = 4$ algorithm of Fig. 4.14. Note that the flow diagram for the small-N algorithm for $N_1 = 3$ and $N_2 = 4$ as well as the WFTA for $N = 12$ follow the same pattern of input additions, followed by multiplications, followed by output additions. In the assimilation of the $N = 4$ flow diagram into the composite diagram for $N = 12$ the multiplication factor of j in the bottom line is carried through to the multiplication sector of the larger algorithm.

In an implementation of the WFTA described by McClellan and Nawab [41] the input mapping and roughly half the additions are carried out in a pre-weave subroutine. This leaves all the multiplications to be carried out as one nested inner loop, which is used recursively. After the multiplications the remainder of the additions and the output mapping are obtained using a post-weave subroutine. Since each allowable transform length has different multiplication factors, an initialization procedure is required to compute these before the multiplication procedure can commence.

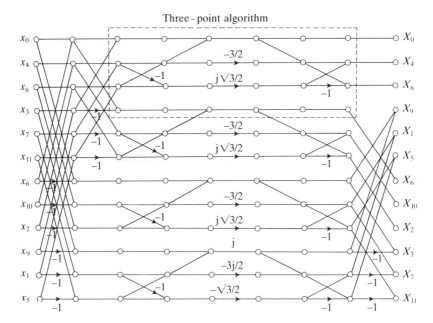

Fig. 4.16 Flow diagram for a Winograd algorithm for $N = 12$.

A number of implementations of the Winograd algorithm and computer programs are described elsewhere. The reader is directed especially to [20], which contains well-documented programs for several of the reduced multiplication algorithms considered in this chapter.

References

1. Algazi, V. R. and Sakrisin, D. J. On the optimality of the Karhunen–Loève expansion. *IEEE Trans. inf. Theory* **IT-15,** 319–21 (1969).
2. Welch, P. D. A fixed point fast Fourier transform error analysis. *IEEE Trans. Audio Electroacoust.* **AU-17**(2), 151–7 (1969).
3. Weinstein, C. J. Roundoff noise in floating point fast Fourier transform computation. *IEEE Trans. Audio Electroacoust.* **AU-17**(3), 209–215, (1969)
4. Zohar, S. Winograd's discrete Fourier transform algorithm. In *Two-dimensional signal processing* (ed. T. S. Huang) Springer, Berlin. pp. 89–160 (1981).
5. Runge, C. and Koenig, G. Die Gründbahren der mathematischer Wissenschaftern. *Vorlesungen über Numerisches Rechnen.* Springer, Berlin (1924).
6. Danielson, G. C. and Lanczon, C. Some improvement in practical Fourier analyses and their application to X-ray scattering from liquids. *J. Franklin Inst.* **233,** 365 (1942).
7. Cooley, J. W. and Tukey, J. W. An algorithm for the machine calculation of Fourier series. *Math. Comput.* **19,** 297 (1965).

8. Good, I. J. The interactive algorithm and practical Fourier series. *J. R. statist. Soc.* **B20,** 361–72 (1958).

9. Gentleman, W. M. and Sande, G. Fast Fourier transforms for fun and profit. *Fall Joint Comput. Conf. AFIPS Proc.* **29,** 563–78. (1966).

10. Morris, L. R. Time efficient radix-4 fast Fourier transform. In *Programs for Digital Signal Processing.* (ed. Signal Processing Committee) IEEE Press, New York 1.8–1 (1979).

11. Bergland, G. D. A fast Fourier transform algorithm using base 8 iterations. *Math. Comput.* **22,** 275–9 (1968).

12. Singleton, R. C. An algorithm for computing the mixed-radix fast Fourier transform. *IEEE Trans. Audio Electroacoust.* **AU-17**(2), 93–100 (1969).

13. Brigham, E. O. *The fast Fourier transform.* Prentice-Hall, Englewood Cliffs, (1974).

14. Winograd, S. On computing the discrete Fourier transform. *Math. Comput.* **32,** 175–99 (1978).

15. McClennan, J. H. and Rader, C. M. *Number theory in digital signal processing.* Prentice-Hall, Englewood Cliffs (1979).

16. Kolba, D. P. and Parks, T. W. A prime factor FFT algorithm using high-speed convolution. *IEEE Trans. Acoust. Speech signal Process.* **ASSP-25,** 90–103 (1977).

17. Nussbaumer, H. J. *Fast Fourier transform and convolution algorithms.* Springer, Berlin (1982).

18. NAG FORTRAN Library Mk. 10. NAG Ltd, Oxford (1983).

19. Hopper, M. J. Harwell Subroutine Library: a catalogue of subroutines. Harwell Report AERE R9185 (1978).

20. *Programs for digital signal processing,* (ed. Signal Processing Committee) IEEE Press, New York, (1979).

21. Dubois, E. and Venetsanopoulis, A. N. A new algorithm for the radix-3 FFT. *IEEE Trans Acoust. Speech signal Process.* **ASSP-26**(3), 222–5 (1978).

22. Theilheimer, F. A matrix version of the fast Fourier transform. *IEEE Trans. Audio Electroacoust.* **AU-17**(2), 158–161 (1969).

23. Saucedo, R. and Schiring, E. E. *Introduction to continuous and digital Control Systems* pp. 570–664. Macmillan, New York (1968).

24. Elliott, D. F. and Rao, K. R. *Fast transforms, algorithms, analysis, applications.* Academic Press, New York (1984).

25. Bergland, G. G. A fast Fourier transform algorithm using base-8 iterations. *Math. Comp.* **22,** 275–9 (1969).

26. Gold, B. and Rader, C. M. *Digital processing of signals.* McGraw-Hill, New York (1969).

27. Polge, R. J., Bhagavan, B. K., and Carswell, J. M. Fast computational algorithms for bit-reversal. *IEEE Trans. Comput.* **C-23**(1), 1–9 (1974).

28. Good, I. J. The relationship between two fast Fourier transforms. *IEEE Trans. Comput.* **C-20**(2), 310–17 (1971).

29. Thomas, L. H. Using a computer to solve problems in physics. *Applications of digital computers.* Ginn, Boston (1963).

30. Winograd, S. Some bilinear forms whose multiplicative complexity depends on the field of constants. IBM Technical Journal, Watson Research Center, New York, Research Report No. RC5669 (1975).

31. Despain, A. M. Very fast Fourier transform algorithm hardware for implementation. *IEEE Trans. Comput.* **C-28**(5), 333–41 (1979).

32. Rader, C. M. Discrete convolution via Mersanne transforms. *IEEE Trans. Acoust. Speech signal Process.* **ASSP-25**(5), 356–9 (1976).

33. Leibowitz, L. M. A simplified binary arithmetic for the Fermat number transform. *IEEE Trans. Acoust. Speech signal Process.* **ASSP-24**(5), 356–9 (1976).

34. Rader, C. M. Discrete Fourier transform when the number of data samples is prime. *IEEE Proc.* **56**, 1107–8 (1968).

35. Pollard, J. M. The fast Fourier transform in a finite field. *Math. Comput.* **25**, 365–4 (1971).

36. McClellan, J. H. Hardware realization of a Fermat number transform, *IEEE Trans. Acoust. Speech signal Process.* **ASSP-24**(3), 216–25 (1976).

37. Sin, W. C. and Constantinides, A. G. On the computation of discrete Fourier transforms using Fermat number transform. *IEEE Proc.* **131F**(1), 7–14 (1984).

38. Burrus, C. S. and Eschenbacker, P. W. An in-place in-order prime factor FFT algorithm. *IEEE Trans. Acoust. Speech signal Process.* **ASSP-29**, 806–16 (1981).

39. Rotweiler, J. H. Implementation of the in-order prime factor transform for variable sizes, *IEEE Trans. Acoust. Speech signal Process.* **ASSP-30**(1), 105–7 (1982).

40. Chu, S. and Burrus, C. S. A prime factor FFT algorithm using distributed arithmetic, *IEEE Trans. Acoust. Speech signal Process* **ASSP-30**(2), 217–27 (1982).

41. McClellan, J. H. and Nawab, H. Complex general-*N* Winograd Fourier transform algorithm WFTA. In *Programs for digital signal processing* pp. 1.7-1–1.7-20. IEEE Press, New York (1979).

OTHER FAST TRANSFORMATIONS

5.1 Introduction

Fast transformation of discrete orthogonal series is not confined to the sinusoidal functions and considerable interest has been given in recent years to the development of a wide spectrum of non-sinusoidal discrete transform algorithms. This is due to their applicability to high-speed digital computation in areas such as image processing, communication systems, signal processors and in many application situations where the microprocessor is seen as an important signal processing tool [1, 2].

The most important of these non-sinusoidal transforms are the Walsh and Haar discrete transforms considered earlier (Chapter 2), and the related hybrid transforms such as the slant and Walsh-derived cosine transform. Because the Walsh functions are binary-related their generation and implementation is fairly simple. Fast algorithms based on sparse matrix factoring have been developed, often similar in form to the FFT. Since these fast Walsh transforms require only additions/subtractions in their evaluation compared with the complex arithmetic of the fast Fourier transform routines they offer considerable economies in computer implementation.

The structures of a number of the non-sinusoidal transform algorithms are sufficiently alike to encourage development of generalized transform methods that can provide a systematic transition from the rectangular-based fast Walsh transform to the sinusoidally-based Fourier transform. Several of these generalized procedures will be considered here.

The chapter concludes with a brief look at methods of implementing fast two-dimensional transforms, particularly for image processing (considered in a later chapter).

5.2 The fast Walsh transform

5.2.1 Derivation of Walsh-function series

One route to the fast Walsh transform (FWT) is from consideration of the composition of the Walsh-function series shown earlier in Fig. 2.2. This can be derived in a number of ways. Well-known methods include algebraic synthesis through recursive relationships, multiplication of a set of square-wave (Rademacher) functions, via Boolean logic and through the use of Hadamard matrices. This last method is particularly relevant to sequency transformation and will now be considered.

Let us look first at a discrete Walsh function matrix, W_N corresponding to the transform matrix B given in eqn (3.47). An $N \times N$ matrix can be obtained quite simply by sampling the set of Walsh functions at N equally spaced intervals. Thus for $N = 8$ and referring to Fig. 2.2 we can derive W_N by inspection as

$$
W_8 = \begin{bmatrix}
1 & 1 & 1 & 1 & 1 & 1 & 1 & 1 \\
1 & 1 & 1 & 1 & -1 & -1 & -1 & -1 \\
1 & 1 & -1 & -1 & -1 & -1 & 1 & 1 \\
1 & 1 & -1 & -1 & 1 & 1 & -1 & -1 \\
1 & -1 & -1 & 1 & 1 & -1 & -1 & 1 \\
1 & -1 & -1 & 1 & -1 & 1 & 1 & -1 \\
1 & -1 & 1 & -1 & -1 & 1 & -1 & 1 \\
1 & -1 & 1 & -1 & 1 & -1 & 1 & -1
\end{bmatrix} = \begin{bmatrix}
\text{WAL}(0, t) \\
\text{WAL}(2, t) \\
\text{WAL}(3, t) \\
\text{WAL}(4, t) \\
\text{WAL}(5, t) \\
\text{WAL}(6, t) \\
\text{WAL}(7, t) \\
\text{WAL}(8, t)
\end{bmatrix}
$$

(5.1)

with the corresponding WAL(n, t) functions shown opposite each row of the matrix.

Now if we consider a Hadamard matrix H_8 derived from eqns (3.55) and (3.56) as

$$
H_8 = H_4 . H_2 = \begin{bmatrix}
1 & 1 & 1 & 1 & 1 & 1 & 1 & 1 \\
1 & -1 & 1 & -1 & 1 & -1 & 1 & -1 \\
1 & 1 & -1 & -1 & 1 & 1 & -1 & -1 \\
1 & -1 & -1 & 1 & 1 & -1 & -1 & 1 \\
1 & 1 & 1 & 1 & -1 & -1 & -1 & -1 \\
1 & -1 & 1 & -1 & -1 & 1 & -1 & 1 \\
1 & 1 & -1 & -1 & -1 & -1 & 1 & 1 \\
1 & -1 & -1 & 1 & -1 & 1 & 1 & -1
\end{bmatrix} = \begin{bmatrix}
\text{HAD}(0, t) \\
\text{HAD}(2, t) \\
\text{HAD}(3, t) \\
\text{HAD}(4, t) \\
\text{HAD}(5, t) \\
\text{HAD}(6, t) \\
\text{HAD}(7, t) \\
\text{HAD}(8, t)
\end{bmatrix}
$$

(5.2)

we are able to obtain a set of N discrete Walsh functions, identical to those shown in eqn (5.1), but arranged in a different order. This is known as Hadamard or Kronecker ordering and is related to the Walsh sequency order by a bit-reversal and a Gray-code conversion [2]. Representing the equivalent sequence of N row values given in eqn (5.2) as a Hadamard series, $\text{HAD}(n, t)$, $n = 0, 1, \ldots, 7$. Then

$$\text{HAD}(n, t) = \text{WAL}(b(u), t) \qquad (5.3)$$

where u represents a bit-reversal for n and $b(u)$ is a Gray-code-to-binary conversion.

This derivation brings out the important fact that the Walsh function series may be expressed in several different ordering sequences, all of which are orthogonal and valid for analysis. There are three ordering conventions in common use: *sequency order* by ascending order of zero crossings, shown in Fig. 2.2 as $\text{WAL}(n, t)$; *natural order,* which results from the Hadamard matrix method of derivation and written as $\text{HAD}(n, t)$; and a third order, the *Dyadic order,* written as $\text{PAL}(n, t)$ and which is the order obtained when the function series is obtained throughout products of Rademacher functions.

Transforms relating to all three ordering conventions for the transformed coefficients are in use. The most economical in terms of program simplicity and speed of calculation are those based on a Hadamard matrix representation in natural order given by eqn (5.2) and are called *fast Hadamard transforms* (FHT).

The discrete Walsh transform, corresponding to eqn (2.51) may be expressed in matrix terms as we did for the Fourier transform of eqn (4.39) as

$$X_n = H_n \cdot x_i \qquad (5.4)$$

(neglecting scaling by N) where x_i is a column vector of order N representing the sampled values of an input signal. As before the transformation is carried out by multiplying the input data vector x_i, with elements of the Hadamard matrix, H_n, and processing down each column in turn summing each set of column products to obtain the transformed output data vector X_n. Since the products represent multiplication by $+1$ or -1 this procedure will be seen to require $N(N-1)$ additions or subtractions.

5.2.2 Fast transformation

We saw earlier that a fast transformation algorithm may be obtained by a process of matrix factorization. This is particularly easy to do with the Hadamard matrix, since it is derived from Kronecker products where it is implicit that N be capable of decomposition into a series of prime numbers. The principles of matrix factorization were discussed in Section 3.3.5 following the procedure originally suggested by Good [3]. Carrying this out for H_8 results in

$$H_8 = \begin{bmatrix}
1 & 1 & 1 & 1 & 1 & 1 & 1 & 1 \\
1 & -1 & 1 & -1 & 1 & -1 & 1 & -1 \\
1 & 1 & -1 & -1 & 1 & 1 & -1 & -1 \\
1 & -1 & -1 & 1 & 1 & -1 & -1 & 1 \\
1 & 1 & 1 & 1 & -1 & -1 & -1 & -1 \\
1 & -1 & 1 & -1 & -1 & 1 & -1 & 1 \\
1 & 1 & -1 & -1 & -1 & -1 & 1 & 1 \\
1 & -1 & -1 & 1 & -1 & 1 & 1 & -1
\end{bmatrix}$$

$$= \begin{bmatrix}
1 & 1 & 0 & 0 & 0 & 0 & 0 & 0 \\
0 & 0 & 1 & 1 & 0 & 0 & 0 & 0 \\
0 & 0 & 0 & 0 & 1 & 1 & 0 & 0 \\
0 & 0 & 0 & 0 & 0 & 0 & 1 & 1 \\
1 & -1 & 0 & 0 & 0 & 0 & 0 & 0 \\
0 & 0 & 1 & -1 & 0 & 0 & 0 & 0 \\
0 & 0 & 0 & 0 & 1 & -1 & 0 & 0 \\
0 & 0 & 0 & 0 & 0 & 0 & 1 & -1
\end{bmatrix}^3 . \qquad (5.5)$$

It is the presence of so many zero components in this factored form for H_n that leads directly to a considerable reduction in the product calculations required for the FWT from (in this case) $N(N-1)$ to $N \log_2 N$ additions and subtractions.

This derivation will give identical nodal column stages and a flow diagram based directly on this is given in Fig. 5.1. This gives an output in natural order and will require auxiliary storage for intermediate stage results. Note that only paired terms are involved and no self-contained butterfly loops are employed. This repeated type of flow diagram is known as a *constant geometry* structure. There are obvious advantages in hardware construction [4, 5] of the transform algorithm particularly for implementation using integrated circuits [6].

The recursive structure of this algorithm may be expressed as follows. Referring to Fig. 5.1 where $N = 2^p$ and designating the coefficient value at each node position as $y_i(n)$ where $i = 0, 1, 2, 3$ and with n as the position down each column of nodes $n = 1, 2, \ldots, 8$ then the input, intermediate, and transformed values are respectively

$$y_0(n) = x(n)$$
$$y_{i+1}(n) = y_i(2n - 1) + y_i(2n) \qquad \text{for } 0 \leqslant n \leqslant N/2$$
$$= y_i(2n - N - 1) - y_i(2n - N) \quad \text{for } N/2 < n \leqslant N. \qquad (5.6)$$

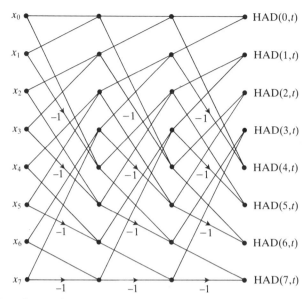

Fig. 5.1 Flow diagram for a constant-geometry FWT.

These results can easily be extended into higher dimensions and programmed accordingly. The program will require a working area of at least $N/2$ locations as well as N places for the data.

The constant geometry algorithms do not permit in-place calculation for the intermediate transform coefficients. As with the radix-2 FFT, a series of fast Walsh transforms are available that are 'in-place' and have a close similarity with the C–T and S–T FFT algorithms. Indeed it is possible to derive a suitable computer program for the FWT by modification of some C–T fast transform algorithms. This is obtained by reducing the trignometric values used in the program to unity and removing the complex part of the operation, since the Walsh transform is a real one. An example is given in the flow diagram for the Walsh–Hadamard fast transform (WHT) in Fig. 5.2 for $N = 8$, which corresponds to the C–T FFT given in Fig. 4.9. Since only addition and subtraction are required to compute the transform coefficients, the algorithm can be considerably faster than the radix-2 FFT.

A difficulty with the WHT is that to obtain a sequency-ordered output, which is desirable for many applications, it is necessary to carry out a bit-reversal on the input or output terms together with a Gray code reordering of the output. A method of avoiding this Gray code reordering was first suggested by Manz [7]. He replaced some of the normal butterflies of the C–T derived algorithm by reversed ones (Fig. 5.3), and further stated a series of rules by which the blocks of butterflies for reversal may be selected. Referring to Fig. 5.2, these are summarized

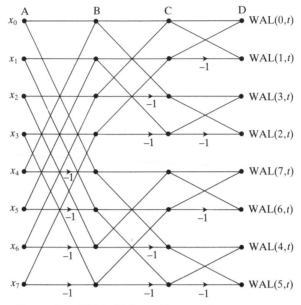

Fig. 5.2 Flow diagram for a Walsh–Hadamard fast transform.

as:

1. Lines originating from the input nodes A are *never* reversed.

2. The next set of nodes between B and C are defined as having a number of 'blocks' of butterflies. A block denotes a group of computations that are disconnected from their neighbours above and below. Thus two such blocks are present between B and C and four between C and D.

3. Reversals occur in blocks. At each column of nodes after the first, the bottom block is reversed. Alternate blocks are then reversed, with the top block always remaining unreversed.

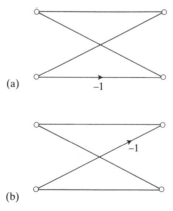

Fig. 5.3 Normal and reversed butterfly diagrams.

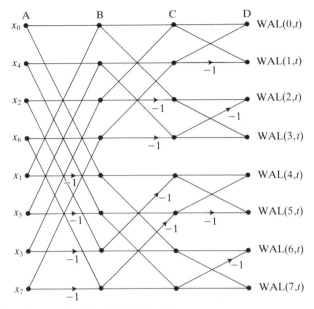

Fig. 5.4 Flow diagram for Manz's FWT algorithm.

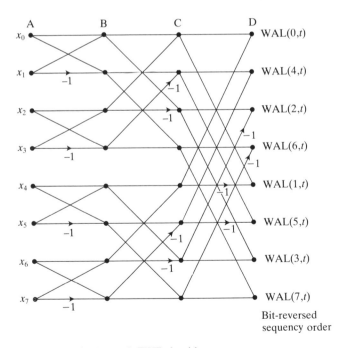

Fig. 5.5 Flow diagram for Larsen's FWT algorithm.

The flow diagram for Manz's algorithm is shown in Fig. 5.4 for $N = 8$. This gives output values in sequency order but the input is in bit-reversed order. A similar algorithm derived from the alternative C–T flow diagram (Fig. 4.8) is due to Larsen [8]. This accepts the data in sequential order but gives the transformed coefficients in bit-reversed sequency order (Fig. 5.5). For those applications requiring a double transformation (in general a Walsh transform acts as its own inverse and a separate inverse transform is not required), the use of the Larsen transform followed by the Manz transform could eliminate the bit inversion stage completely. In this sense the two transforms could be regarded as complementary to each other.

A number of alternative FWT algorithms have been developed. Some are able to accept normally ordered input data and provide an output in sequency order at the cost of increased program complexity and additional memory storage. Others are designed to give an output in a particular form required for specific applications. Two examples of these are the Rademacher–Walsh transform used in logic design [9] and the CAL–SAL ordering used in signal processing [10].

5.2.3 Phase-invariant transforms

A problem that arises in the practical use of the Walsh transform, noted in Section 2.4.3, is that it is not invariant with phase shifts of the input signal as is the case with Fourier transformation. This characteristic can be shown to inhibit the interpretation of results in such application areas as pattern recognition and image processing. A shift-invariant transform, known as the *rapid transform,* (RT) has been described by Reitboeck and Brody to overcome this limitation [11]. Although the transformed data is independent of cyclic shifts of the input signal, some reduction in the range of information concerning the sequency distribution of the upper range of coefficients is found and the transform does not provide its own inverse. The invariance is obtained quite simply by replacing the subtractive terms obtained stage by stage at each node of the fast transform algorithm by their absolute values. It is necessary however to choose an algorithm in which the sample pairs are obtained from coefficients spaced $N/2$ values apart, such as Manz's algorithm shown in Fig. 5.4. Alternative algorithms for the computation of the RT have been described by Ulman [12] and Kunt [13] and a general range of shift-invariant transforms is given by Wagh [14].

5.3   The fast Haar transform

The set of discrete Haar functions for $N = 8$ shown in Fig. 2.3 can be sampled at N equally spaced intervals along the time base to obtain the

Haar matrix, Ha_8 as,

$$
Ha_8 = \begin{bmatrix}
1 & 1 & 1 & 1 & 1 & 1 & 1 & 1 \\
1 & 1 & 1 & 1 & -1 & -1 & -1 & -1 \\
\sqrt{2} & \sqrt{2} & -\sqrt{2} & -\sqrt{2} & 0 & 0 & 0 & 0 \\
0 & 0 & 0 & 0 & \sqrt{2} & \sqrt{2} & -\sqrt{2} & -\sqrt{2} \\
2 & -2 & 0 & 0 & 0 & 0 & 0 & 0 \\
0 & 0 & 2 & -2 & 0 & 0 & 0 & 0 \\
0 & 0 & 0 & 0 & 2 & -2 & 0 & 0 \\
0 & 0 & 0 & 0 & 0 & 0 & 2 & -2
\end{bmatrix}
\begin{matrix}
\left.\vphantom{\begin{matrix}1\\1\end{matrix}}\right\}A \\
\left.\vphantom{\begin{matrix}1\\1\end{matrix}}\right\}B \\
\\
\left.\vphantom{\begin{matrix}1\\1\\1\\1\end{matrix}}\right\}C
\end{matrix} .
$$

$$(5.7)$$

The discrete Haar transform can then be stated in terms of the matrix equation,

$$X_n = Ha_N . x_i \qquad (5.8)$$

(again neglecting scaling by $1/N$). As with other unitary matrices the matrix for Ha_N can be factored to obtain the fast Haar transform (HT) algorithm. Since the matrix is not a symmetrical one a separate algorithm is required for the direct and the inverse transform.

Algorithms to compute the HT and its inverse have been developed by Andrews [15], and flow diagrams for these are given in Fig. 5.6 (a and b) for $N = 8$. Multiplication of the sum/difference values by 1 or 2 is indicated in the diagram. It will be seen that at each step in the calculation (other than the first) half the points require no further calculation. (The multiplications can all be delayed until the transformation is completed). Thus the total number of additions/subtractions is

$$N + N/2 + N/4 + , \ldots , + 2 = 2(N - 1). \qquad (5.9)$$

Transformation time for this algorithm is therefore linearly proportional to the number of terms N, in contrast to the FWT or the radix-2 FFT where it is proportional to $N \log_2 N$.

Haar matrices and hence their sparse matrix factors contain irrational numbers (i.e. powers of $\sqrt{2}$). A version of the Haar transform, known as the *rationalized Haar transform,* has been introduced by Lynch and Reis [16]; this removes the irrational numbers and replaces them with integer powers of two. This version preserves all the properties of HT and is more suitable for hardware implementation.

From Fig. 5.6 it is apparent that this algorithm does not follow the general form of the C–T FFT algorithms, in which each stage can consist of repetitions of butterfly subsections. As a step towards the development of a common algorithm for all three fast transformations, FFT, FWT, and

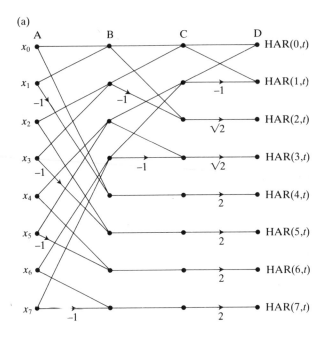

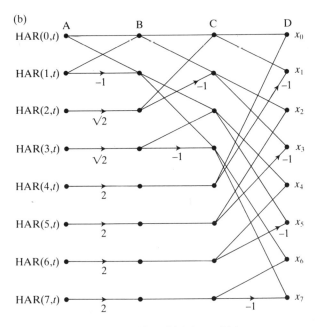

Fig. 5.6 Flow diagram for the HT algorithm. (a) Direct; (b) inverse.

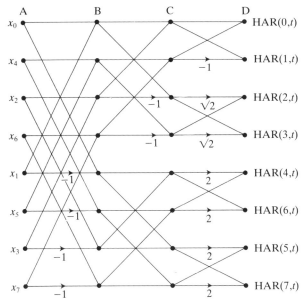

A   B   C   D

Fig. 5.7 Flow diagram for Ahmed's Haar transform.

HT, a C–T version of the HT has been described by Ahmed *et al.* [17]. The columns of the $\textbf{\textit{Ha}}_8$ matrix given in eqn (5.7) are rearranged using successive bit-reversals acting sequentially on subsections of the matrix shown as A, B, and C in eqn (5.7). The result is the matrix given in eqn (5.10), which can be formulated as the signal flow diagram shown in Fig. 5.7, i.e.

$$
\textbf{\textit{Ha}}_8' = \begin{bmatrix}
1 & 1 & 1 & 1 & 1 & 1 & 1 & 1 \\
1 & -1 & 1 & -1 & 1 & -1 & 1 & -1 \\
\sqrt{2} & 0 & -\sqrt{2} & 0 & \sqrt{2} & 0 & -\sqrt{2} & 0 \\
0 & \sqrt{2} & 0 & -\sqrt{2} & 0 & \sqrt{2} & 0 & -\sqrt{2} \\
2 & 0 & 0 & 0 & -2 & 0 & 0 & 0 \\
0 & 2 & 0 & 0 & 0 & -2 & 0 & 0 \\
0 & 0 & 2 & 0 & 0 & 0 & -2 & 0 \\
0 & 0 & 0 & 2 & 0 & 0 & 0 & -2
\end{bmatrix}. \quad (5.10)
$$

In order to achieve this simplification and to obtain an in-place Haar transform it is necessary to carry out a bit-reversal on the position of the input signal values and on the output coefficients and this will add to the time needed to carry out this transformation.

5.4   The fast slant transform

The slant transform was first used by Enomoto and Shibata [18] primarily as a method of bandwidth compression in television transmission systems. Although the importance of this transform has been overshadowed by the discovery of the cosine transform, considered later, it still plays an important role where a relationship is sought among different orthogonal transforms. The discrete slant series is a hybrid formation in which a discrete Walsh function series and a discrete series based on a ramp or slant waveform are combined to provide an orthogonal series, described as $SLA(n, t)$ whose characteristics match those found in individual scanned lines of a television image. A continuous set of slant functions for $N = 8$ is shown in Fig. 5.8.

The *fast slant transform* (FST) may be derived from matrix considerations commencing as with the Hadamard–Walsh transform, with the Hadamard matrix for $N = 2$, i.e.

$$H_2 = \begin{bmatrix} 1 & 1 \\ 1 & -1 \end{bmatrix}$$

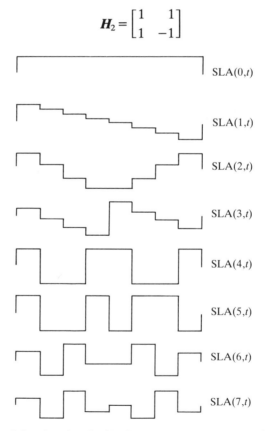

SLA(0,t)

SLA(1,t)

SLA(2,t)

SLA(3,t)

SLA(4,t)

SLA(5,t)

SLA(6,t)

SLA(7,t)

Fig. 5.8   A set of slant functions for $N = 8$.

which is taken as the first slant matrix S_2 of $N \times N$ terms (ignoring scaling factors). The slant transform matrix for $N = 4$ can be written as

$$S_4 = \begin{bmatrix} 1 & 1 & 1 & 1 \\ a+b & a-b & -a+b & -a-b \\ 1 & -1 & -1 & 1 \\ a-b & -a-b & a+b & -a+b \end{bmatrix} \qquad (5.11)$$

where a and b are real constants whose values are dependent on two factors:

1. S_4 must be orthogonal;
2. the step size of the slant basis rows must be uniform throughout its length.

The first condition is achieved by making $b = 1/5$, hence $a = 2/5$ and the second condition achieved by making $a = 2b$. This leads to a matrix for S_4 as,

$$S_4 = \begin{bmatrix} 1 & 1 & 1 & 1 \\ 3/5 & 1/5 & -1/5 & -3/5 \\ 1 & -1 & -1 & 1 \\ 1/5 & -3/5 & 3/5 & -1/5 \end{bmatrix}. \qquad (5.12)$$

A general expression may be developed for a slant matrix of order N in terms of a matrix of order $N/2$ in a similar manner to Hadamard matrix expansion through Kronecker multiplication; the value of the a and b constants being determined by a recursive relationship [19]. The slant matrix so formed will possess a sequency property with a number of its middle rows being identical to a sequency ordered Hadamard matrix. The similarity between the general form of the slant series and the sequency-ordered Walsh series is seen if Fig. 5.8 and Fig. 2.2 are compared.

The slant matrix is used to define the slant transform as,

$$X_n = S_n . x_i. \qquad (5.13)$$

where S_n is a generalized form of the 4×4 matrix S_4 given by eqn (5.12). The matrix is decomposed into a product of a series of sparse matrices to obtain the fast slant transform algorithm, which for $N = 4$ is given as,

$$S_4 = \begin{bmatrix} 1 & 0 & 0 & 0 \\ 0 & 3/5 & 0 & 0 \\ 0 & 0 & 1 & 0 \\ 0 & 0 & 0 & 3/5 \end{bmatrix} . \begin{bmatrix} 1 & 1 & 0 & 0 \\ 0 & 0 & 1 & 1/3 \\ 1 & -1 & 0 & 0 \\ 0 & 0 & 1/3 & -1 \end{bmatrix} . \begin{bmatrix} 1 & 0 & 0 & 1 \\ 0 & 1 & 1 & 0 \\ 1 & 0 & 0 & -1 \\ 0 & 1 & -1 & 0 \end{bmatrix}. \qquad (5.14)$$

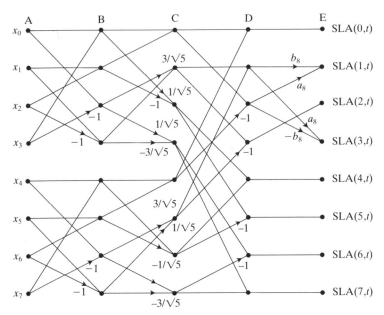

Fig. 5.9 Flow diagram for a FST algorithm.

A flow diagram for $N = 8$ is shown in Fig. 5.9. Note that this algorithm requires $p + 1$ stages ($N = 2^p$) rather than p and is consequently slower than the Walsh transform. For example where $N = 4$ the slant transform requires eight additions/subtractions and six multiplications whereas the corresponding Walsh fast transform requires only eight additions/subtractions.

A hybrid slant transform based on the Haar series is described by Fino [20] in which elements of the Haar matrix are combined with slant functions. Mathematical relationships are stated linking the slant (Walsh) transform with the slant (Haar) transform.

The Walsh FST has been shown to have a good mean-square error coding performance for digital images, almost as good as the ideal Karhunan–Loève transform (KLT) with the advantage of a fast transform algorithm that the KLT does not possess.

5.5 The generalized transform

The close similarities between the algorithmic form of the FFT and FWT has led to the formalization of a general transform algorithm through the work of Ahmed and Rao [21, 22]. This provides a family of $\log_2 N$ transforms including the FFT and FWT at either end of the range. However apart from these and a complex Walsh transform [23] the other orthogonal transforms in the family appear to be only of mathematical interest.

The generalized transform family are defined in terms of a *generalized matrix*

$$\boldsymbol{G}_r(p) = \prod_{k=1}^{p} \boldsymbol{D}_r(p)^k \qquad (5.15)$$

$(r = 0, 1, 2, \ldots, p - 1; \; K = 1, 2, \ldots, p)$ where $\boldsymbol{D}_r^k$ are a set of block diagonal matrices defined as

$$\boldsymbol{D}_r^k(p) = \mathrm{diag}(\boldsymbol{A}_0^r(k), \boldsymbol{A}_1^r(k) \ldots \boldsymbol{A}_{2p-k}^r(k)\}. \qquad (5.16)$$

These block matrices are generated recursively and follow the form,

$$\boldsymbol{A}_m^r(1) = \begin{cases} \begin{bmatrix} 1 & W^{mb} \\ 1 & -W^{mb} \end{bmatrix} & \text{for } m = 0, 1, \ldots, 2^r - 1 \\[3mm] \begin{bmatrix} 1 & 1 \\ 1 & -1 \end{bmatrix} & \text{for } m = 2^r, 2^r + 1, \ldots, 2^{p-k} \end{cases} \qquad (5.17)$$

$$\boldsymbol{A}_m(k)^r = \boldsymbol{A}_m(1)^r \otimes \boldsymbol{I}_2(k - 1) \qquad (5.18)$$

m_b is the bit-reversed equivalent of m.

From the form of these block matrices we see that the generalized matrix must result in a transform algorithm consisting entirely of in-place butterflies. This is shown for $N = 8$ in Fig. 5.10. The transform obtained

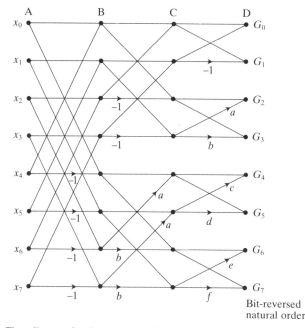

Fig. 5.10   Flow diagram for the generalized transform algorithm.

Table 5.1 Coefficients for the generalized transform

	FWT	Complex BIFORE	FFT
a	1	$-j$	W^2
b	-1	j	$-W^2$
c	1	1	W
d	-1	-1	$-W$
e	1	1	W^3
f	-1	-1	$-W^3$

in this way is dependent entirely on the choice of the coefficients a to f shown in the diagram which are in turn, dependent on the value of the single coefficient r that can take the following values:

$r = 0$ yields the FWT with the output in natural order and this derivation is known as the BIFORE transform [24, 25];

$r = 1$ yields the complex BIFORE transform [23] which finds a use where the input takes the form of a complex data series;

$r = p - 1$ yields a FFT with the output in bit-reversed order corresponding to the C–T algorithm given in Fig. 4.8

where r lies between $r = 1$ and $r = p - 1$ a further set of $p - 3$ orthogonal algorithms are generated.

Values of the coefficients, a_1 to a_6 for the three cases of interest are given in Table 5.1.

A more recent attempt to find common ground in the formulation of an algorithm for fast orthogonal transforms is described by Fino and Algazi [26]. This concerns the group of discrete unitary and orthogonal transform matrices defined by eqn (3.37) which we can express through eqn (3.18), as

$$C \cdot [C^*]^T = I \qquad (5.19)$$

where I is a unit matrix of the same order as C, i.e. $N \times N$. The unitary transformation is then expressed as,

$$X_n(u) = C_u \cdot x_i \qquad (5.20)$$

where C_u is the uth unitary transform matrix.

The group of u unitary transform matrices include the discrete Fourier, Walsh/Hadamard, Haar, slant, and several others.

Previous development of C–T and S–T FFT, FWT, and generalized transform algorithms approached the design problem by analytic means, generally through the Good decomposition into sparse matrices [3]. With Fino's technique the approach is synthetic where a few types of fast unitary matrices of small order are used to generate recursively fast unitary transforms of arbitrary order. In this respect the method resembles Winograd's derivation of large transform matrices through the product of small-N matrices and indeed an analytical form of the Winograd algorithm in this way has been suggested by Vlasenko and Rao [27].

The basic ideas in the synthetic approach are:

1. The permutation of columns through a block diagonal matrix.
2. The permutation of rows through the *perfect shuffle* algorithm.
3. Synthesis of the complete transform matrix through Kronecker products of a set of standard matrices.

A unitary matrix is defined as

$$C = P^{\mathrm{T}}[\mathrm{diag}\,A]P[\mathrm{diag}\,B] \qquad (5.21)$$

where $[\mathrm{diag}\,A]$ and $[\mathrm{diag}\,B]$ are block diagonal matrices identical in form but having different sets of coefficient values for each of the unitary matricrs, and P and P^{T} are perfect shuffle permutation matrices.

5.5.1 The perfect shuffle

This is a constant geometry algorithm describing a set of node interconnections for fast transform algorithms by which the number of interconnections may be halved. This was first suggested by Stone in connection with parallel processing hardware for the FFT [28]. He noted that given an active node that carries out simultaneously both summation and subtraction of pairs of input values to produce sum and difference terms, only a rearrangement of input values is needed to reduce the number of interconnection values to $N/2$. The resulting interconnection arrangement is called the *perfect shuffle*, by analogy with the shuffling of two packs of m cards each $(a, b, c, \ldots, m$ and $a', b', c', \ldots, m'$ into $a, a', b, b', c, c', \ldots, m, m')$. A form of this constant geometry transformation for a single stage is shown in Fig. 5.11 for $N = 8$. We can also define this interconnection as the mapping of the indices on the left of the diagram onto those on the right according to a permutation P_s such that

$$\begin{aligned} P_s(i) &= 2i & 0 \leqslant i \leqslant N/2 - 1 \\ &= 2i + 1 - N & N/2 < i \leqslant N - 1. \end{aligned} \qquad (5.22)$$

A useful feature of the perfect shuffle is that if the indices are represented in binary form then the ith element may be shuffled to its new position i' by cyclically rotating the bits one position to the left. Thus for,

$$i = i_{n-1}2^{n-1} + i_{n-2}2^{n-2} + \ldots + i_1 2 + i_0$$
$$i' = i_{n-2}2^{n-1} + i_{n-3}2^{n-2} + \ldots + i_0 2 + i_{n-1} \qquad (5.23)$$

which is seen to agree with eqn (5.22).

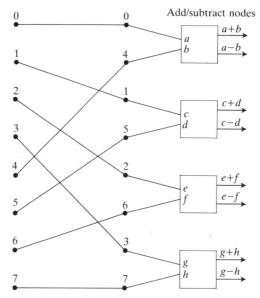

Fig. 5.11   A perfect shuffle algorithm.

5.5.2   Fast unitary transforms

The unitary matrix defined in eqn (5.21) is applied by Fino in a recursive expression to generate a family of fast unitary transforms sharing a common computational algorithm. This may be expressed for a radix-2 Fourier transform, for example, in the simplified form

$$X_n = PF_k \otimes X_{n-1} \qquad (5.24)$$

where F_k is a butterfly block diagonal matrix of the form

$$\begin{bmatrix} 1 & W^k \\ 1 & -W^k \end{bmatrix} \qquad (5.25)$$

and X_{n-1} is the vector output of the preceding computational stage and P is a perfect shuffle matrix.

Repeated application of eqn (5.24) results in a block matrix structure leading to a flow diagram, which is shown for $N = 8$ in Fig. 5.12. This corresponds to the decimation-in-time C–T FFT algorithm given in Fig. 4.8, although requiring six stages instead of three. The individual stages consist of Stone's perfect shuffle stage, three butterfly stages, and two bit-reversal stages.

Using combinations of these basic stage configurations, Vlasenko has shown that the FWT, FFT, HT, FST, and several other orthogonal and non-orthogonal fast transform algorithms can be carried out [27, 29]. The

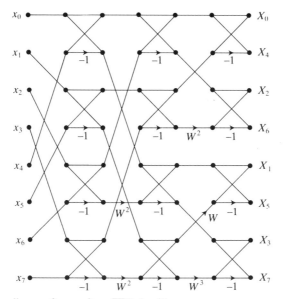

Fig. 5.12 Flow diagram for a unitary FFT algorithm.

common organization of such transform algorithms can lead to considerable simplification in the computer programs, particular for hardware implementation using parallel processors.

5.6 Two-dimensional transformation

A general procedure exists for extending the single-dimensional fast transformation into multiple dimensions. This was introduced in Section 2.5 for two-dimensional image transformation using the DFT. In order to consider this in matrix terms the eqns (2.59) and (2.60) are restated below as eqns (5.25) and (5.26)

$$X_{n,m} = 1/MN \sum_{i=0}^{N-1} \sum_{k=0}^{M-1} x_{i,k} \cdot W_N^{ni} W_M^{mk}. \tag{5.25}$$

This is the discrete transformation, where

$$x_{i,k} = x(nT, mS)TS$$

and T and S are the sampling periods in the x and y direction, respectively.

The exponentials are replaced by

$$W_N = \exp(-j2\pi/N)$$

and

$$W_M = \exp(-j2\pi/M). \tag{5.26}$$

The inverse transformation can similarly be expressed as

$$x_{i,k} = \sum_{n=0}^{N-1} \sum_{m=0}^{M-1} x_{n,m} W_N^{-ni} W_M^{-mk}. \tag{5.27}$$

The two-dimensional data is in the form of a $N \times M$ matrix $x_{i,k}$, i.e.

$$x_{i,k} = \begin{bmatrix} x(0, 0)x(0, 1), \ldots, x(0, N-1) \\ x(1, 0)x(1, 1), \ldots, x(0, N-1) \\ \vdots \qquad \vdots \qquad\qquad \vdots \\ x(M-1, 0)x(M-1, 1), \ldots, x(M-1, N-1) \end{bmatrix}. \tag{5.28}$$

Equation (5.25) can be evaluated through a one-dimensional DFT by partitioning into the following form,

$$X_{n,m} = \frac{1}{N} \sum_{i=0}^{N-1} W^{ni} \frac{1}{M} \sum_{k=0}^{M-1} x_{i,k} W^{mk}. \tag{5.29}$$

The inner summation is calculated first and is seen to be the DFT of each column of the data matrix $x_{i,k}$ and we can write

$$X_{n,k} = \frac{1}{M} \sum_{k=0}^{M-1} x_{i,k} W^{m,k}. \tag{5.30}$$

The coefficients, $X_{n,k}$ can be expressed as an $N \times M$ data matrix

$$X_{n,i} = \begin{bmatrix} X(0, 0)X(0, 1), \ldots, X(0, N-1) \\ X(1, 0)X(1, 1), \ldots, X(0, N-1) \\ \vdots \quad \vdots \qquad\qquad \vdots \\ X(M-1, 0)X(M-1, 1), \ldots, X(M-1, N-1) \end{bmatrix}. \tag{5.31}$$

A transformation of this data matrix results in

$$x_{i,k} = \frac{1}{N} \sum_{i=0}^{N-1} X_{n,k} W^{ni}. \tag{5.32}$$

This implies that the final transform coefficients, $x_{i,k}$ are formed by taking the DFT of each row of $X_{n,k}$.

Thus the calculation of the two-dimensional transform can proceed in two steps:

1. A separate DFT is taken for each of the N columns of data. These will give N sets of M coefficient values arranged in the form of a matrix, $X_{n,k}$.

2. Each of the M rows of matrix $X_{n,k}$ are then transformed separately to give M sets of N coefficient values to form the final transformed matrix, $X_{n,m}$. The process is illustrated in Fig. 5.13 (from the author's earlier publication [2]) for $N \times M = 8 \times 8$.

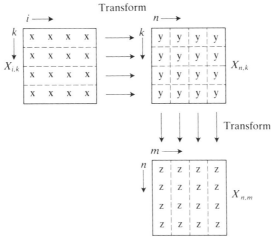

Fig. 5.13 Two-dimensional transformation. (From Beauchamp [2].)

This approach is often called the *row-column method*. An improved method which reduces the number of real operations at the expense of a more complex algorithm is the polynomial transform due to Rader and Brenner [30]. This is described in some detail and compared with other polynomial transforms in [31]. The row-column method can be applied to other fast orthogonal transformations such as the Walsh or Haar. For $N = M$ then $2N^2 \log_2 N$ arithmetic operations are required, compared with N^4 for the direct evaluation of eqn (5.25).

An alternative method due to Bates [32] also uses a single-dimensional transformation to derive a two-dimensional transform. This rearranges an image matrix into a one-dimensional vector with the columns following one another sequentially. A one-dimensional transform of N^2 values is taken of the vector and the resulting output rearranged again into matrix form.

5.6.1 The fast cosine transform

The discrete cosine transform, introduced in Section 2.5.1 is of increasing importance in image processing, where its performance has shown to be near optimal [33]. As indicated in eqn (2.65), a *fast cosine transform* (FCT) may be evaluated through a $2N$ point fast Fourier transform by adding N zeros to the data to be transformed. Other operations such as multiplication by $\exp(-\jmath n/2N)$ and taking the real part of the result are also required. This method requires double the data storage of the single FFT, and although some organizational improvements are possible [34], using the FFT directly as a route to the FCT is not very efficient and other methods are to be preferred.

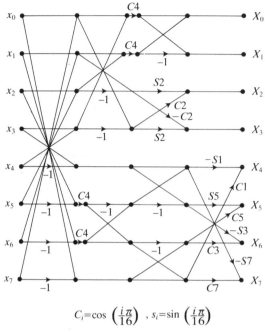

$$C_i = \cos\left(\frac{i\pi}{16}\right), \quad s_i = \sin\left(\frac{i\pi}{16}\right)$$

Fig. 5.14 Flow diagram for the FCT algorithm.

One of the most successful of these is the algorithm due to Chen, Smith, and Fralich [35]. A flow diagram is given in Fig. 5.14. This will be seen to consist of alternating cosine/sine butterflies with binary matrices inserted to reorder the matrix elements into a form that preserves a bit-reversed pattern at every other node. It is thus an in-place algorithm and apart from normalization can be reversible and used as an inverse transform. Extension of the flow diagram to the next power of two involves adding a set of normal butterflies and a series of alternative cosine/sine butterflies to yield a new set of odd transform coefficients. In matrix terms the transform matrix for this FCT takes the form

$$\boldsymbol{C} \cdot \boldsymbol{T}_N = \boldsymbol{P}_N \cdot \left[\begin{array}{c|c} CT_{N/2} & 0 \\ \hline 0 & R_{N/2} \end{array} \right] \cdot \boldsymbol{I}_N \tag{5.33}$$

where $\boldsymbol{P}_N$ is an $N \times N$ permutation matrix translating the transformed vector from bit-reversed order to natural order and $\boldsymbol{I}_N$ is a unit matrix. $\boldsymbol{C} \cdot \boldsymbol{T}_{N/2}$ is the previous lower order of the cosine matrix for $N/2$ and $R_{N/2}$ is a constants matrix defined in [35]. The algorithm is about six times faster than that of the double FFT algorithm.

A later development by Narasimha and Peterson [36] showed that almost the same improvement in speed is obtainable using a standard radix-2 FFT, where N can be an even number. In this algorithm the data is reordered such that two real transforms can be evaluated in one pass through a complex N-point FFT (see Section 4.2.5). Using the fast Winograd algorithm for the FFT, which does not constrain the data to be complex, can also lead to an efficient DCT at the cost of program complexity and more elaborate data reordering routines [37].

A quite different approach is to obtain the FCT via the FWT, which can reduce the number of non-integer multiplications required. This may be seen if eqn (2.63) is considered in matrix form, i.e.

$$X_n = CT_n \cdot x_i \qquad (5.34)$$

where CT_n is an $N \times N$ cosine matrix.

Evaluating CT_n for $N = 8$ obtains

$$CT_8 = \begin{bmatrix} 0.354 & 0.354 & 0.354 & 0.354 & 0.354 & 0.354 & 0.354 & 0.354 \\ 0.490 & 0.416 & 0.278 & 0.098 & -0.098 & -0.278 & -0.416 & -0.490 \\ 0.462 & 0.191 & -0.191 & -0.462 & -0.462 & -0.191 & 0.191 & 0.462 \\ 0.416 & 0.098 & -0.490 & -0.278 & 0.278 & 0.490 & -0.098 & -0.416 \\ 0.354 & -0.354 & -0.354 & 0.354 & 0.354 & -0.354 & -0.354 & 0.354 \\ 0.278 & -0.490 & -0.098 & 0.416 & -0.416 & 0.098 & 0.490 & -0.278 \\ 0.191 & -0.462 & 0.462 & -0.191 & -0.191 & 0.462 & -0.462 & 0.191 \\ 0.098 & -0.278 & 0.416 & -0.490 & 0.490 & -0.416 & 0.278 & -0.098 \end{bmatrix}.$$

$$(5.35)$$

From this we can see a recognizable symmetry in the signs of the cosine matrix elements. There is, in fact, a one-to-one correspondence between the signs of these terms and those of the Walsh matrix W_8 shown in eqn (5.1). This indicates that the basic vectors of the DCT are essentially *amplitude-modulated* versions of the basic vectors of the FWT and this provides a rationale for the development of a conversion algorithm between the two transformations. It has been shown elsewhere that the link between W_N and CT_N takes the form of a transformation matrix having a diagonal block structure containing many zero value terms [38]. The sparse nature of this matrix enables a fast cosine transform to be obtained and, despite the need first to compute the FWT, the number of additions/subtractions is also reduced.

A similar method is described by Ghanbari and Pearson [39] and applied to the implementation of a television image compression system. In order to reduce the number of arithmetic elements needed in this implementation the Hadamard ordering (e.g. H_8 of eqn (5.2)) is taken as the sign-comparative Walsh transform and the columns and rows of the DCT matrix are rearranged to conform with this vector basis. The advantage of this method is that with a suitable decomposition algorithm the adder and subtractor and associated delay circuits in a hardware realization can be shared between all the samples in an 8×8 sub-matrix of elements and a single shift-in-place of the input data produces all four pairs of data simultaneously for subsequent arithmetic operation.

References

1. Yuen, C. K., Beauchamp, K. G., and Robinson, G. P. S. *Microprocessor systems and their application to signal processing.* Academic Press, London (1982).
2. Beauchamp, K. G. *Applications of Walsh and related functions.* Academic Press, New York (1984).
3. Good, I. J. The interaction algorithm and practical Fourier analysis. *J. R. statist. Soc.* **B20,** 361–72 (1958); **B22,** 372–5 (1960).
4. Muniappan, K. and Kitai, R. Walsh spectrum measurement in natural, dyadic and sequency ordering. *IEEE Trans. electromag. Compat.* **EMC-24** (1), 46–9 (1982).
5. Carl, J. W. and Swartwood, R. V. A hybrid Walsh transform computer. *IEEE Trans. Comput.* **C-22**(7), 669–72 (1973).
6. Smith, E. G. Computer-aided design of a PLA implemented fast Walsh–Hadamard transform device. *IEEE Proceedings of the Fifth International Conference on Pattern Recognition,* pp. 183–91 (1980).
7. Manz, J. W. A sequency-ordered fast Walsh transform. *IEEE Trans. Audio Electroacoust.* **AV-20**3), 204–5 (1972).
8. Larsen, H. An algorithm to compute the sequency-ordered Walsh transform, *IEEE Trans. Acoust. Speech Signal Process.* **ASSP-24,** 335–6 (1976).
9. Hurst, S. L., Miller, D. M., and Muzio, J. C. *Spectral techniques in digital logic.* Academic Press, New York (1984).
10. Kunt, M. In-plane computation of the Hadamard transform in CAL–SAL order. *Signal Process.* (Lausanne) **1**(3), 327–31 (1979).
11. Reitboeck, H. and Brody, T. P. A transformation with invariance under cyclic permutation for application in pattern recognition. Westinghouse Research Laboratory Scientific Paper No. 68, F1-ADAPT-P1 (1968).
12. Ulman, L. J. Computation of the Hadamard transform and the *R*-transform in ordered form. *IEEE Trans. Comput.* **C-19,** 359–60 (1970).
13. Kunt, M. On the computation of the Hadamard transform and the *R* transform in ordered form, *IEEE Trans. Comput.* **C-24,** 1120–21 (1975).
14. Wagh, M. D. and Kanetkar, S. V. A class of translation-invariant waveforms, *IEEE Trans. Acoust. Speech Signal Process.* **ASSP-25,** 203–5 (1977).
15. Andrews, H. C. and Caspari, K. L. A generalized technique for spectral analysis. *IEEE Trans. Comput.* **C-19,** 16–25 (1970).
16. Lynch, R. T. and Reis, J. J. Haar transform image coding, *Proceedings of the National Telecommunications Conference,* Dallas, pp. 44.3-1–44.3-5 (1976).
17. Ahmed, N., Natarajan, T., and Rao, K. R. Cooley–Tukey type algorithm for the Haar transform. *Electron. Lett.* **9,** 276–8 (1973).
18. Enomoti, H. and Shibata, K. Orthogonal transform coding system for television signals. *Proceedings of a Symposium on Applied Walsh Functions,* Washington DC AD727000, pp. 11–17 (1971).
19. Pratt, W. K., Welch, L. R., and Chen, W. H. Slant transform for image coding, *Proceedings of a Symposium on Applied Walsh Functions,* Washington DC AD744650, pp. 229–34 (1972).

20. Fino, B. J. and Algazi, V. R. Slant–Haar transform. *IEEE Proc.* **62,** 653–4 (1974).
21. Ahmed, N., Rao, K. R., and Schultz, R. B. A generalized discrete transform. *IEEE Proc.* (*Lett.*) **59,** 1360–2 (1971).
22. Rao, K. R., Mrig, L. C. and Ahmed, N. A modified generalized discrete transform *IEEE Proc.* **61,** 668–9 (1973).
23. Ahmed, N. and Rao, K. R. Complex Bifore transform. *Electron. Lett.* **6**(8), 256–8 (1970).
24. Ohnsorg, F. Binary Fourier representation. Spectrum Analysis Technique Symposium, Honeywell Research Center, Hopkins, Minn., September (1966).
25. Ahmed, N., Rao, K. R., and Abdussattar, A. L. BIFORE or Hadamard transform. *IEEE Trans. Audio Electroacoust.* **AU-19**(3), 225–34 (1971).
26. Fino, B. J. and Algazi, V. R. A unified treatment of fast unitary transforms. *SIAM J. Comput.* **6**(4), 700–17 (1977).
27. Vlasenko, V. and Rao, K. R. A generalized approach to orthogonal transform algorithms. *IEEE 21st Midwest Symp. Ccts. and Syst.* Ames, pp. 378–83 (1978).
28. Stone, H. S. Parallel processing with the perfect shuffle. *IEEE Trans. Comput.* **C-20**(2), 153–61 (1971).
29. Vlasenko, V. and Rao, K. R. Unified matrix treatment of discrete transforms. *IEEE Trans. Comput.* **C-28**(12), 934–8 (1979).
30. Rader, C. M. and Brenner, N. M. A new principle for fast Fourier transformation. *IEEE Trans. Acoust. Speech Signal Process.* **ASSP-24,** 264–5 (1976).
31. Nussbaumer, H. J. *Fast Fourier transform and convolution algorithms.* Springer, Berlin (1982).
32. Bates, R. M. Multi-dimensional BIFORE transform. Ph.D. Dissertation, Kansas State University (1971).
33. Ahmed, N., Natarajan, T., and Rao, K. R. Discrete cosine transform. *IEEE Trans. Comput.* **C-23,** 90–3 (1974).
34. Haralick, R. M. A storage efficient way to implement the discrete cosine transform. *IEEE Trans. Comput.* **C-25,** 764–5 (1976).
35. Chen, W. H., Smith, C. H., and Fralick, S. S. A fast computational algorithm for the discrete cosine transform. *IEEE Trans. Commun.* **COM-25,** 1004–9 (1977).
36. Narasimha, M. J. and Peterson, A. M. On the computation of the discrete cosine transform. *IEEE Trans. Commun.* **COM-26,** 934–6 (1978).
37. Narasimha, M. J. and Peterson, A. M. Design of a 24-channel transmultiplexer. *IEEE Trans. Acoust. Speech Signal Process.* **ASSP-27**(6), 752–62 (1979).
38. Hein, D. and Ahmed, N. On a real-time Walsh/Hadamard/cosine transform image processor. *IEEE Trans. electromagr. Compat.* **EMC-20,** 453–7 (1978).
39. Ghanbari, M. and Pearson, D. E. Fast cosine transform implementation for television signals, *IEE Proc.* **F129**(1), 59–68 (1982).

6

IMPLEMENTATION

6.1 Introduction

While many applications of transformation algorithms will be incorporated into programs running on 'stand-alone' microprocessors or main-frame time-shared computers a number of tasks require individual hardware/software solutions either because of the processing speed or data volume involved (e.g. real-time radar signal processing or digital image processing) or because of their working environment (e.g. machine tool control or vision robotics). In addition the widespread availability of the dedicated microprocessor and specialized digital processing systems such as the programmable logic array may make more attractive a dedicated design for a single task than by considering this task as one of a number to be carried out sequentially by the overall system.

In this chapter consideration will be given to hardware implementation of digital transformation, considered either as a self-contained operation, or as part of a signal processing system. Several new and interesting developments have contributed to the design of such devices in the last decade. These include techniques such as bit-slice operation and pipe-lining, and component design using charge-coupled devices or surface-acoustic wave applications. Further, the continuing reduction in memory costs are leading to the development of new FFT algorithms in which the arithmetic and logic operations are reduced by the use of very large memory devices to hold the pre-calculated constants and to act as data accumulators [1]. Above all the considerable development in very large-scale integration (VLSI) has given rise to the programmable signal processor (PSP) and more recently the 'single-chip' PSP—devices that are able to carry out domain transformation allied with arithmetic operations, which may be applied to a number of user-programmed signal processing tasks.

The earliest of these hardware designs were the various attempts to produce a hard-wired transformation unit. It will be helpful in considering the later more successful devices if the basic ideas behind these hard-wired transformers are first examined.

6.2 Hardware transformation

Real-time transformation using software methods and time-shared computers or even stand-alone microcomputers can be achieved without too much difficulty for signals having bandwidths up to several hundred Hz. This is however insufficient for many important applications such as radar or speech processing; for these applications, high-performance dedicated methods are needed.

The special-purpose hardware and techniques necessary to overcome the speed and memory limitations of the general purpose computer may be considered with reference to the algorithmic procedure of the FFT. We saw earlier that software implementation for the FFT is at its simplest with radix-2 C–T or S–T algorithms where the only transformation procedure used is the two-node butterfly (Fig. 4.11). This is also true for hardware implementation, but here, since we are likely to be concerned with speed, the sequential routing of pairs of data values through a *single hardware section* would be very slow. If we consider τ as the time to complete one butterfly operation then, using a C–T fast transformation, $\tau n 2^{n-1}$ seconds would be needed for a 2^n point transformation (neglecting intermediate-stage storage operations). A *parallel approach* will be faster but will require n butterflies per stage, and each of the stage interconnections will be different. If a constant-geometry algorithm is used however, a single-stage hardware can be employed and 2^n sets of data routed n times through the one stage. Transformation will be achieved in approximately τn seconds. The fastest arrangement of all is a *pipeline system* whereby all n stages, each processing 2^n sets of data, are operated simultaneously. After the first n transformations a data transformation rate of one complete transformation for each γ, i.e. the time to complete one butterfly, will be achieved. This requires the maximum amount of hardware, namely $n 2^n$ butterflies.

Designs for the fastest processing speed will employ parallel or pipeline methods and integrated circuits in their realization. One promising method is to associate a fast microprocessor with a small but very fast *array processing attachment*, which takes over the highly repetitive array processing computations together with some of the critical memory managements tasks, leaving the microprocessor free for efficient supervision and control of the transformation process [2].

A general technique that may applied to both parallel and pipeline methods is to use very efficient *bit-slice architecture* which effectively introduces parallel processing operations on each sample of a digitized signal.

6.2.1 Bit-slice operation

Let us look first then at these small-scale parallel processing operations and the use that can be made of them. Bit-slice devices are modular components, generally processing four bits only and having the control structures necessary to link to other similar structures 'built-in' to the bit-slice device [3]. Several such devices can be operated simultaneously to produce a subsystem capable of processing a much bigger word-length and at much the same speed as a single device processing a four-bit word. Special operations such as digital filtering and multiplication can be implemented faster and hence more efficiently than many single-chip microprocessors at the present time—a factor of between five and ten times has been quoted [4], with a digital filtering sampling speed in the region of 1 million samples s^{-1} [5]. This improvement in multiplication speed is valuable in domain transformation and the bit-slice device has found wide use in fast Fourier and Walsh hardware implementation [6–8].

It is salutory however, to point out here that the speed and capability of the VLSI single-chip processor is increasing and the advantage gained by bit-slice operations involving separate bit-slice elements is unlikely to last the present decade.

A 4-bit-slice arithmetic element has been described by Magar and is briefly described below as an example of the implementation problems that need to be overcome in designing a unit of this type. Magar's element is designed to carry out the multiplication and adding/subtraction operations required in a basic FFT butterfly [7]. This incorporates a multiplier, and independent adder/subtractor, four peripheral registers and four data multiplexors. These are all contained on a single chip mounted in a 40-pin package. A functional schematic diagram of the device is shown in Fig. 6.1 (included by courtesy of Professor Magar and Academic Press).

The multiplicand input register R_1 can be loaded via two input ports selected by multiplexer control, thus facilitating easy operation in multibus systems. The multiplier word is externally serialized and applied to the y input port. The adder can be accessed externally via registers R_2 and R_3 multiplexed to the Z and X_2 input ports, respectively. Results from the chip are obtained at the Q output port from register R_4, or alternatively fed back to the adder input register R_2 via an internal feedback path, thus assisting the accumulation operation. Control of the chip is achieved via two quasi-independent clocked instruction sets. These instructions are loaded at the I_1 and I_2 inputs to the decoders under the control of the clocked inputs CK_1 and CK_2, respectively. A number of bit-slice arithmetic elements are needed to realize a full butterfly structure having the required data word length. The complete arithmetic device would be used as an additive element for a microprocessor system programmed to carry out the FFT algorithm. Using parallel techniques of this type, FFT processors with data throughput rates in excess of 10 MHz have been configured.

A commercial bit-slice device used in fast parallel transform implementation is the Advanced Micro Devices AM2901. This is a four-bit

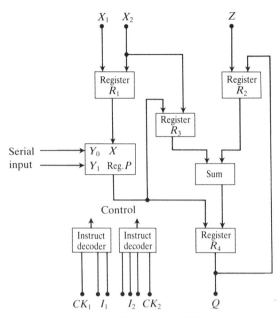

Fig. 6.1 A bit-slice arithmetic element (After Magar [7]).

processor device for which a simplified schematic diagram is given in Fig. 6.2 (courtesy of Advanced Micro Devices Ltd). The AM2901 is a small bipolar microprocessor consisting of a 16 word by 4-bit read-only memory (RAM), a high speed arithmetic and logic unit and a 16-word scratch pad memory. Several registers are included to permit shifting, encoding and multiplying operations. With its 9-bit microinstruction word the AM2901 is a microprogrammable device behaving as a small computer, having the facility to coordinate its operations along a common data bus with other similar devices [7]. Design of bit-slice processing devices and their software support is given in [2, 7, 9, 10].

A successful application of the AM2901 is the single-board FFT computer described by Schirm [11]. This carries out a fixed-point FFT ($N = 1024$) at a speed of 50 kHz using a multi-bus parallel operation of five separate AM2901 processors and one fast arithmetic unit. It achieves a fast working speed through the parallel operation of a number of separate processors arranged to handle all the organization and data combinations and by using a very fast arithmetic unit to process sequentially all the butterfly transform operations required.

6.2.2 Parallel processing

More usually, parallel operation implies the use of a number of butterfly units operating simultaneously. Special-purpose hardware designs based

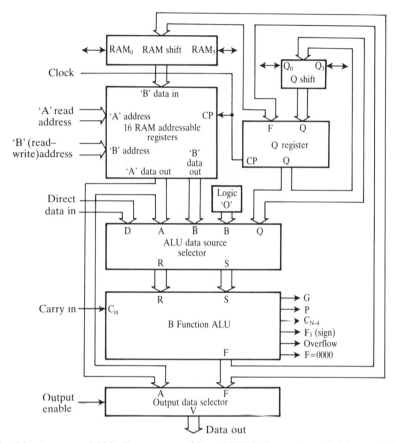

Fig. 6.2 A commercial bit-slice processor (The AM2901). (by courtesy of Advanced Micro Devices Ltd.)

directly on the algorithmic form given by the flow diagram of a fast transformation have been employed for a number of years to produce fast parallel transformations [12]. Those based on the in-place C–T or S–T algorithm consist of a set of standard butterfly modules, a storage module, and other supporting modules, all controlled by a high-speed processor.

A typical butterfly module is shown in Fig. 6.3. Yuen, Beauchamp and Robinson [13]. This is a byte-slice device (i.e. $N = 8$), the performance of which can be extended for an arithmetic precision of 16, 24, or 32 bits by interconnecting several such modules. Each module contains multipliers and adder/subtractors. A RAM would be included for complex data storage and a PROM to hold the 'twiddle factor' constants (sine and cosine pairs) used in the FFT calculation. The data input to the butterfly

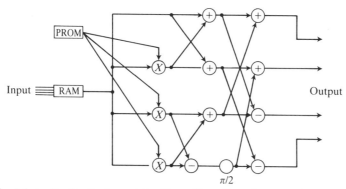

Fig. 6.3 A byte-sliced butterfly module. (From Yuen, Beauchamp, and Robinson [13].)

module consists of sets of four complex numbers supplied simultaneously either from a storage module or from another butterfly module. These inputs are stored in the data RAM and, when a block of data has been assembled in the RAM, groups of four complex values will be read and the radix-4 butterfly operation carried out. The four new complex numbers created by the butterfly operation can be output to another butterfly or storage module or may be returned to the RAM of the butterfly module used in the next stage of the FFT calculation [13].

The design of the butterfly module is considerably simpler for the Walsh transform, since this does not need to handle complex numbers. Several designs are described in [14].

A disadvantage with the hardware design of all these butterfly processors is that each stage in the transformation process is different, so that for a signal having $N = 2^p$ values, p different sets of interconnections need to be fabricated. As described in Chapter 4, an alternative process for the matrix factorization can lead to the constant-geometry algorithm. Two alternative factorization routines are possible. In one the input data can be processed in adjacent pairs of samples or where the sample pairs are separated by $N/2$ samples. This is shown for the Hadamard matrix in Fig. 6.4, where the factorization can be expressed (for $N = 8$) as

$$H_8 = PPP = QQQ. \tag{6.1}$$

The first of these, PPP, results in a constant-geometry flow diagram in which adjacent pairs of data elements are chosen for addition or subtraction and stored in consecutive locations for subsequent processing (see Fig. 5.1). The second, QQQ, involves fetching pairs of data elements separated by $N/2$ locations and after adding or subtracting, storing the results in adjacent locations for subsequent processing. In either case, since the interconnection pattern is the same for all stages of the computation, the hardware can be reduced to one stage by feeding

$$P = \begin{bmatrix} 1 & 1 & 0 & 0 & 0 & 0 & 0 & 0 \\ 0 & 0 & 1 & 1 & 0 & 0 & 0 & 0 \\ 0 & 0 & 0 & 0 & 1 & 1 & 0 & 0 \\ 0 & 0 & 0 & 0 & 0 & 0 & 1 & 1 \\ 1 & -1 & 0 & 0 & 0 & 0 & 0 & 0 \\ 0 & 0 & 1 & -1 & 0 & 0 & 0 & 0 \\ 0 & 0 & 0 & 0 & 1 & -1 & 0 & 0 \\ 0 & 0 & 0 & 0 & 0 & 0 & 1 & -1 \end{bmatrix} \qquad Q = \begin{bmatrix} 1 & 0 & 0 & 0 & 1 & 0 & 0 & 0 \\ 1 & 0 & 0 & 0 & -1 & 0 & 0 & 0 \\ 0 & 1 & 0 & 0 & 0 & 1 & 0 & 0 \\ 0 & 1 & 0 & 0 & 0 & -1 & 0 & 0 \\ 0 & 0 & 1 & 0 & 0 & 0 & 1 & 0 \\ 0 & 0 & 1 & 0 & 0 & 0 & -1 & 0 \\ 0 & 0 & 0 & 1 & 0 & 0 & 0 & 1 \\ 0 & 0 & 0 & 1 & 0 & 0 & 0 & -1 \end{bmatrix}$$

Fig. 6.4 Two forms of constant geometry matrix.

the output back to the input and repeating the operation p times. This is shown in Fig. 6.5 for one stage in the process. A number of constant-geometry systems for Hadamard–Walsh transforms are described in [11, 15, 16].

Some simplification of the hardware implementation of these algorithms may be obtained if a combined adder/subtractor unit replaces alternate nodes in each stage calculation. This is the *perfect shuffle* described earlier (in Section 5.5.1). In an adaption by Pease [17] it is demonstrated that the perfect shuffle interconnection is all that is required to obtain a parallel FFT. This implementation applies the perfect shuffle algorithm shown in Fig. 5.11 in a recursive process, so that

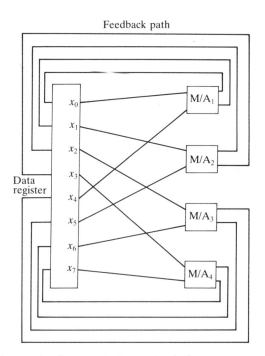

Fig. 6.5 Parallel processing for a constant geometry device.

the two weighted sums are produced simultaneously and may be fed back into the input register to replace the original values. p iterations of the data are required and the final transformed values is obtained in bit-reversed binary order.

A similar arrangement is described by Shirata and Nakatsuyama [18], which uses a combination of perfect shuffle interconnections and a set of adder/subtractor units to obtain fast Walsh transformation in either natural, dyadic, or sequency order. The shuffle interconnections in this implementation are not, however, uniform and six stages are required for $N = 16$, together with a set of eight adder/subtractor units. Control of the output from these units enables the selection of intermediate outputs at each stage in the iteration so that the output can be obtained in the selected order. A simpler design, also based on the perfect shuffle, is described in [19]. This provides the selection of any one of the three orderings through the use of a set of selection circuits following the adder/subtractor units. A considerable amount of symmetry is achieved with this design, which simplifies its implementation in hard-wire or LSI form.

6.2.3 Pipeline processing

In a pipeline processor a number of identical butterfly modules are connected in series, each performing one stage of the FFT (or other) algorithm before passing the results to the next module. Each set of cascaded modules is called a pipeline [20].

A pipeline processor is illustrated in Fig. 6.6 for $N = 8$. Three butterfly modules are required and two sets of intermediate memory and reordering logic. In the timing diagram for these three stages the dashes represent butterfly operations performed on data from the preceeding N-point transform, while the plus signs indicate butterfly operations carried out in the following N-point transform. From this we see that three sets of output data pairs from the first butterfly module need to be stored before the second butterfly module is activated to commence the calculation of the second of the three stages in the calculation. Finally, on clock pulse five, all three butterfly stages are in operation and a stream of transformed data pairs are produced. Thus the first set of data from this final stage in the pipeline is available after five clock pulses. This is the fixed initial delay of the pipeline, after which the data flow is at the rate of operation for a single butterfly stage.

Two pipelines of modules can be arranged to process each half of the complex FFT inputs, thus permitting a complete set of N data values to be input in $N/2$ clock pulses. This division into two parallel data streams is not entirely separate, however, since each FFT output is a function of *all* the inputs. It is necessary therefore to arrange to exchange some of

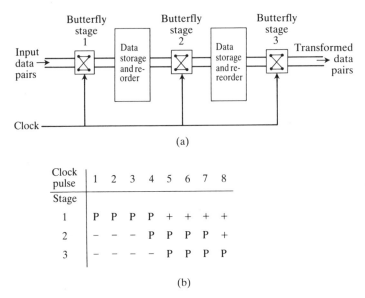

Fig. 6.6   A pipeline processor.

the inputs of the two pipelines at one point along the parallel route. Additional speed of operation through the operation of four or more pipelines with consequent input interconnection can also be obtained. In all cases, further modules are required to buffer the data and to insert multiplication by fixed constants (e.g. the 'twiddle factor' constants).

Pipeline processors have been described by Groginsky and by O'Leary [21, 22]. These early developments used discrete component modules and achieved a FFT throughput rate of over 3Mb s^{-1} with $N = 1024$.

While the pipeline approach forms probably the fastest organizational solution to the problem of domain transformation, it has not been widely accepted. This is due in part to the complex hardware needs, despite the high degree of modularity that may be achieved, and to problems of control and signal-to-noise ratio in the transmission of data through the extended system logic. The situation is changing however with the development of the integrated circuit PSP, and several of the programmable signal processors described in Section 6.5 use pipeline techniques.

The design problem has been somewhat easier for the non-sinusoidal transforms. A structure for the hardware realization of the fast Walsh transform has been proposed by Ashouri and Constantinides [23]. The structure uses the basic delay line method employed by Groginsky and Works [21] in their realization of the FFT. A single stage in this method is shown in Fig. 6.7, which corresponds to a single butterfly operation (multiplication by the twiddle factor is not required and the process of

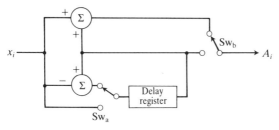

Fig. 6.7 A delay line butterfly implementation.

multiplication of element values is replaced by addition/subtraction). The signal samples are presented sequentially to a pair of summing circuits arranged to add or subtract the current sample with a previous sample held in the delay register. Each register will need to hold $N-1$ sequential samples. Switching of the output connection to obtain the sum or difference of appropriate pairs of samples is made coincident with the switching of the delay register to the correct summer input. Consideration of the complete transform flow diagram (see Fig. 5.2) shows that different switching frequencies will be required for each of the p stages in the algorithm. The control mechanism for the switches is, in fact, a simple synchronized binary counter in which the binary number generated identifies the order of the sample currently being produced at the output. Thus the first transformed output sample will appear at the processor output as the Nth data sample is fed to the input of the processor, i.e. after all the registers are filled. Thereafter processing will occur at the rate of one complete transformed sample every clock period, which is determined solely by the operational time of one butterfly stage.

6.3 Microprocessor implementation

The flexibility and availability of the general-purpose microprocessor makes this attractive as a tool for carrying out the fast transform algorithm. Software implementation is however, very slow, and for an 8-bit word microprocessor the transformation time can take several seconds for $N > 512$. This may be adequate for data acquisition systems or very low frequency applications such as seismic signal analysis, and a number of such implementations have been described [24, 25].

There are in fact a number of limitations of the microprocessor as a domain transformer apart from its slow speed. Programming in assembler or machine code is essential to achieve any reasonable performance and the small word length and limited memory of the microprocessor present some difficulties. One of these concerns data overflow. Since the fast transformation proceeds in a series of stages, the magnitude of numbers in the input sequence can increase by one bit at each stage. After a few stages have been carried out an overflow of the data registers can occur. The problem is overcome by carrying out some form of data scaling; this has to be designed carefully to avoid round-off errors. It is, in any case, one further operation to include in a microprocessor transform implementation.

A radix-2 C–T FFT is described by Luk [24] using a Motorola 6800 8-bit microprocessor which obtains a total transformation time of 318 mS for $N = 64$. Some programming simplification is achieved by Kobylinski using a very modest hardware attachment although this results in a slightly slower transform time than Luk's implementation [26]. In Kobylinski's application a single hardware butterfly logic unit is entered recursively from the controlling 8-bit microprocessor so as to carry out sequentially all the in-place butterfly operations required in a radix-2 C–T FFT algorithm.

More extensive hardware attachments operating in a parallel or pipeline mode offer considerably improved performance. Examples are the bit-slice arithmetic modules, described in the previous section and the fast programmable logic and other array multipliers considered in [5, 6, 27]. In these examples the microprocessor is used as a control to handle data transfers, input/output data movement, bit-inversion, and other data re-shuffling routines, leaving arithmetic operations to be carried out in the special-purpose attachment. Using a microprogrammable arithmetic device of this sort the data throughput rate is controlled entirely by the multiplication time of the special-purpose device and it becomes possible to compute a transformation within a few ms. [6].

The complexity of the complete hardware transformation system however increases considerably with N, so that special hard-wired designs are limited to a very few applications. Instead hardware transform development has proceeded towards the programmable signal processor (PSP), which can support a number of alternative or additional signal processing operations and not simply transformation. PSPs will be considered in Section 6.5.

In recent years the basic capability of the general-purpose microprocessor has increased with the introduction of high-speed 16/32-bit devices, and this has improved the performance of software transformation for the FFT and other transforms. This has occurred at the same time as the newer faster algorithms have become available with their requirements for small-N transformations as the building blocks for larger-N transformations—this also favours microprocessor implementation [28].

The Motorola MC68000, a fast 16/32-bit microprocessor, appears particularly suitable for this role. Applying this device to the FFT Gibson and McCabe, compare the performance obtainable with the conventional

Table 6.1 Performance of a general purpose MC68000 microprocessor programmed as a FFT device; $N = 15/16$ (after Gibson and McCabe [30].)

	Radix 2 C–T (MS)	WFTA (MS)	PFA (MS)
Arithmetic	2.072	0.709	0.975
Data transfers	0.708	0.750	1.039
Overheads	0.424	0.095	1.503
Total transform time	3.204	1.554	3.517
Program size (Memory: byte)	330	682	600

radix-2 C–T algorithm, the WFTA, and a version of the PFA described by Despain [29]. This is summarized in Table 6.1 taken from their results [30]. (With acknowledgement to R. M. Gibson, D. P. McCabe and the IEEE).

Clearly the increased speed for the architecture of the MC68000 with its clock rate of 8 MHz enables a considerable improvement in transformation rate over the earlier 8-bit microprocessors to be realized. Further the results show that the use of small-N algorithms for the WFTA when programmed for the microprocessor can be advantageous in enabling faster transformation than the conventional C–T FFT, although at the cost of more than twice the memory requirements. The lower arithmetic requirements of Despain's algorithm (where data shifts replace multiplication) appear to be nullified by an increase in operating overheads in this comparison.

The ratios between performance levels between the three algorithms is dependent on the prime factors chosen for a given N size, as discussed in Chapter 4, so that any comparisons of this kind can only be taken as a general guide. In another example of the use of the MC68000 for fast Fourier transformations, particularly for large N values, Said and Dimond obtained rather faster times [31]. They obtained a transformation time of 48 ms for a 256-point transform using a radix-2 C–T FFT algorithm with complex data input. The transformation time for $N = 16$ is given as about half that in the previous example.

Optimization of FFT algorithms for a microprocessor application is very dependent on the processor architecture. Using the MC68000, the time taken to execute data operations within the memory is longer than when internal registers are used. For this reason it is fairly common to execute the butterfly routines using only the internal registers. Said and Dimond additionally make a distinction between those twiddle factors having values of ± 1 and $\pm j$ and those having other values. This enables the multiplication operation for the 1s and js to be changed to an add instruction during the butterfly execution period. A reduction in the time needed for data scaling (see earlier) is also proposed. By recognizing that overflow does not occur until several stages of the calculation have been carried out, it is possible to defer scaling until this stage is reached.

Microprocessor implementation for the FWT is faster than the FFT and so enables useful results to be obtained using the cheaper 8-bit processors. Durgan and Lai [32] have described the Walsh transformation of a low-frequency signal ($f = 50\,\text{Hz}$) in a real-time situation using an Intel 8080 microprocessor. Also Collado and colleagues describe a Motorola 6809 system for calculating the fast Haar transform at sample frequencies up to $7\,\text{kHz}$ [33]. In the Durgan and Lai implementation the in-place algorithm of Fig. 5.4 is used with the output obtained in bit-reversed order. Here $N = 1024$ and the software includes provision for generating the PSD from the Walsh spectrum using the periodogram definition given by eqn (2.57). Addition of separate hardware can improve the transformation speed and in the case of the non-sinusoidal signals this need not be too elaborate. In a design due to Muniappan and Kitai [34], the Walsh transformation is carried out using Berauer's algorithm [35]. This is implemented directly in hardware logic and is an interesting example of the use of *direct memory access* (DMA) to obtain fast data transfer.

6.4 Microtechnology

6.4.1 VLSI transformation

Despite the availability of cheap microprocessors, digital logic, and a wealth of sophisticated fast transform algorithms, there remains a need for a cost-effective transformation device that can handle a large-N data input at a speed high enough to be of use in on-line situations. The key to this development lies with a single-chip transform, which can be used on its own or as a component substrate in a general-purpose signal processing chip.

A major difficulty in deriving a suitable VLSI design is a practical one of obtaining an optimum area–time performance in which a maximum signal throughput (i.e. minimum transform time) may be achieved without using up too much valuable area on the chip surface [36, 37]. This is an extension of the argument for determining a cost–performance measure that relates to the number of transforms produced per unit time and to the cost of the components and interconnection [38]. The actual number of active elements on a chip may be less important therefore than the way that they are interconnected [39].

The various possibilities for chip layout follow the choice of a single shared system, a parallel approach, or pipe lining. There are certain simplistic advantages in defining a single multiply–add cell (equivalent to a single butterfly) and to arrange to carry out N^2 sequential operations on this to yield a DFT. However weighting constants would also need to be calculated on the chip and the entire concept can be shown to require a disproportunate chip area for its realization [40]. Parallel operation on a linear array of N multiply–add cells is suggested in [41] and [42], in which recursive operation on p sets of such cells $(N - 2^p)$ will produce the complete DFT, again by direct methods. An equivalent pipeline solution makes use of N^2 multiply–add cells so that the data being processed spends just one computational step in each row of cells before moving on to the next row. After $4N - 3$ steps the transformed data is available at the output and continues to arrive at a rate governed by the processing time of a single cell. Through suitable arangement of interconnection lengths between the matrix of cells a good compromise between area and time can be realized.

These are all direct transform methods. As with the software DFT an increase in efficiency can be expected if FFT algorithm methods are employed. One obvious way is to provide one multiply–add cell for each node in the FFT flow diagram. The inter-cell connections can provide a problem in layout design and additional wiring is necessary to ensure that neither input nor output is in a shuffled order [43]. The cell interconnections become simpler if the perfect shuffle form of algorithm is employed, thus reducing the number of multiply–add cells required (although at the same time this will reduce the speed of operation). Other interconnections are feasible (nine possibilities are listed in [40]) and the perfect shuffle appears to be a good candidate for optimum performance in single-chip design.

Where the VLSI transform configuration is realized using several interconnected chips, attention is directed towards minimizing the number of off-chip interconnections, since these result in much slower data exchange than can occur within the chip [44]. The design scheduling algorithms that result from this approach favour the C-T type of flow diagram where this consists of groups of butterflies carrying out sub-processing tasks complete within themselves rather than the apparently faster but unsymmetrical FFT algorithm (compare the symmetrical FFT algorithm of Fig. 4.9 with the unsymmetrical Winograd algorithm of Fig. 4.16). A single-chip Walsh transform does not require multiplication by a range of constants and gives less problem in design simulation. A preliminary study has been carried out by Nick [45], directed towards the production of a single-chip Walsh function spectral analyser. The proposed design includes registers storing the vector products representing Walsh function values, and a set of line drivers for onward transmission of the data series. Using a fairly conventional logic circuit family (actually only AND–OR gates) the transformation speed is given as about 150 kHz for $N = 64$. A *programmable logic array* (PLA) design for a single-chip FWT has also been described by Smith [46].

6.4.2   Charge-coupled devices

Charge-coupled devices (CCD) offer a number of very attractive advantages for digital processing due to their essential simplicity, cheapness, and combined analog–digital operation [47]. The device is essentially a shift register formed on a silicon substrate as a string of closely spaced solid-state capacitors. A CCD can store and transfer analog charge signals introduced electrically or optically into the substrate. It can be considered as the equivalent to an analog tapped delay line having digital control—a combination that results in a useful signal-processing tool combining the best features of digital and analog techniques.

Fig. 6.8 illustrates the principle of the CCD arranged as a transversal filter. An analog input signal is applied to a series of M interconnecting charge-storage devices acting as delay stages D, each of which retains the value of the input at the end of a clock period. The signal is non-destructively sampled, the stage output potentials multiplied by weighting coefficients h_k, $(k = 0, 1, 2, \ldots, M - 1)$ and the products summed. Very large numbers of delay stages are possible in a single device and a 500-stage CCD is quite usual. This makes it valuable for image processing and a cost-effective method for achieving matched filtering or convolution. It is important to note that sampled analog signals are processed under digital control, which eliminates the requirements for analog–digital conversion. The limitations of a CCD lie in its dynamic range, linear charge transfer efficiency, and leakage. These disadvantages are outweighed in many applications by the low cost of the CCD and its small size.

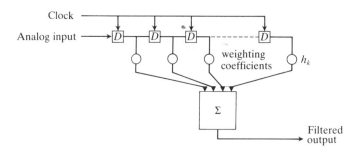

Fig. 6.8 A CCD transversal filter.

The CCD finds its main application in digital filtering and in spectral analysis. As a transformation device it is particularly effective in providing significant potential saving over digital fast transformation within the limitations noted earlier. The way in which the CCD is used to implement a discrete transform is however, quite different from the conventional hardware constructions considered earlier. In the digital fast transform algorithms a major aim is to reduce the number of multiplications. This factor is no longer important with the CCD since transversal filters can be constructed to carry out large numbers of multiplications simultaneously in real time. Instead we look for algorithms in which the major part of the computation is carried out by the transversal filter. Most of the CCD implementations of the DFT make use of the *chirp-Z transform algorithm* (CZT). As applied to the CCD this has some performance limitations compared with the digital FFT and is somewhat less flexible, but it has considerable advantages in its low-cost manufacture, small size, and good reliability [48].

A definition of the DFT was given in eqn (4.2). If the substitution of

$$2ni = n^2 + i^2 - (n - i)^2 \tag{6.2}$$

is made and W^{in} is expressed as an exponential, the transform equation looks like

$$X'_n = \underbrace{\exp(-j\pi n^2/N)}_{C} \sum_{i=0}^{N-1} (\underbrace{x_i \exp(-j\pi i^2/N)}_{A})\underbrace{\exp(j\pi(n - i)^2/N}_{B}. \tag{6.3}$$

The three operations that comprise the CZT algorithm are shown as A, B, and C in this factored equation and correspond to the three sections of the schematic diagram shown in Fig. 6.9. The signal x'_i is multiplied with the chirp signal, expressed by the exponential term $-j\pi i^2/N$, passed through a matched chirp filter (see later) and convolved with a second chirp signal, $-j\pi n^2/N$. The signals are actually sampled analog signals distinguished here from digital sampled signals by the prime, '.

To implement the N-point CZT algorithm, a CCD filter of length $2N - 1$ is arranged as a chirp filter [49]. The signal to the filter is

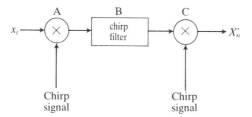

Fig. 6.9 Simplified diagram for a CCD chirp-Z Fourier transformer.

pre-multiplied by a swept frequency (chirp) waveform and post-multiplied by a similar chirp waveform in order to obtain the proper phasing of the DFT coefficients. After the pre-multiplication phase the input needs to be blanked for N clock periods while the convolution with the chirp filter takes place. The post-multiplication can be avoided if the PSD and not simply the DFT is required.

This approach for Fourier transformation has the disadvantage of requiring filters of length $2N$ samples for an N-point transform and is inefficient, since only half of the CCD filter contains useful data at any one time. A modification of the CZT, called the *sliding-CZT*, can avoid this blanking stage when the PSD is evaluated [49]. The difference between the sliding CZT and the CZT based directly on the DFT is that in the former the data are shifted by one sample each time a spectral component is calculated. For a simple transformation this would destroy the phase information but with a PSD evaluation this does not contain any phase information and a useful PSD is realized. Fourier transformation can be obtained at rates up to 100 kHz for a 500-stage filter.

The CCD has also been used to implement Rader's prime factor algorithm for the FFT (see Chapter 4) and is described by Jack *et al.* [50]. Some advantages in hardware design are claimed for this method with no significant change in performance or accuracy. A discrete cosine transform using the CCD is given by Kapur *et al.* [51]. A CCD method that makes use of the C–T FFT algorithm directly is also given by Roberts [52]. Here a tapped CCD delay line is connected at each node point to a resistor network matrix used in the formation of a set of scalar products for W_{in} (see Fig. 4.1). The method is only suitable for small values of N however, since N^2 resistors are required and a large array size would be prohibitively expensive to implement.

The CCD has been applied to the Walsh transform by Yarlagadda and Hershey [53]. This employs a version of the constant-geometry algorithm shown in Fig. 5.1. The CCD delay line is used as a serial register to hold a set of $N/2$ adjacent pairs of input samples. Each pair is added (or subtracted) in a single analog summing device, which is multiplexed between pairs of the CCD taps. This occurs automatically as the data traverses through the register past the two summer inputs. The summed values are returned to the input of the CCD delay line from the summer for the next stage in the process. After p stages the completely transformed data begins to emerge from the output of the delay line to result in a bit-reversed Walsh transformation in serial form. Other applications for Walsh transformation are described in [14].

6.4.3 Surface-acoustic-wave devices

Surface-acoustic-wave devices (SAW) form yet another solid-state processing technique that is now well established for frequency filtering and transformation, based on the ability of the device to store energy information on a slowly propagating acoustic wave. There is some similarity and overlapping of functions with the charge-coupled devices introduced in the preceding section. However with bandwidths of several hundred MHz and dynamic ranges extending towards 100 dB, passive SAW devices form a strongly competitive signal processing element [54].

SAW devices are produced by the same photolithographic processing used for metallization in IC manufacturing and are thus compact, reliable, and economical. Their principle of operation is shown in Fig. 6.10. An electric field is generated when an alternating voltage (the signal) is applied to a pair of interlaced metallic combs (known as 'fingers') printed on the surface of piezo-electric material. This field propagates along the material towards a second similar transducer where an inverse piezo-electric effect generates a voltage between the pairs of fingers which constitute the reception transducer. A very high proportion of the wave energy is transmitted in this way without loss and the velocity of propagation is high, typically $3000 \, \text{ms}^{-1}$. Relatively large delays can be achieved for very small path lengths, so that a compact transversal filter can be created having a very wide bandwidth and used for domain transformation.

A common method of implementing the SAW Fourier transform is through the chirp-Z transform described earlier [55, 56]. Other implementations have been described using the sliding CZT transform [57] and the prime factor algorithm [58].

In a similar manner to the CCD CZT, the SAW CZT is implemented through pre-multiplication of an input signal with a chirp waveform, a convolution in a chirp filter, and subsequent post-multiplication with another chirp filter to yield the Fourier transformation as indicated in eqn (6.3).

The chirp waveforms are generated by impulsing physically realizable SAW chirp filters. These are reproduced through a *dispersive delay line*. A dispersive delay line is a filter whose delay is a function of the input

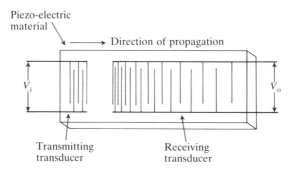

Fig. 6.10 A SAW device.

frequencies. Since in SAW devices the delay is proportional to the length of the propagation path, the fabrication of a chirp filter may be obtained simply by using a long dispersive transducer having gradually increasing separation between the electrode fingers for the receiving transducer and a short non-dispersive transducer at the transmitting end. The receiving end transducer shown in Fig. 6.10 is of this type. A number of such filter designs are described in the literature [59]. It has been shown [56] that the inverse Fourier transform can be obtained using a similar configuration of chirp filters. A chirp-Z SAW filter is very similar in operation to a CCD filter and application of the two methods are often considered as alternatives [60].

The two-level characteristics of the Hadamard matrix are particularly suitable for implementation in a piezo-electric substrate. Several designs have been suggested whereby a series of simple comb detectors are printed on a substrate as illustrated in Fig. 6.11. All of these detectors receive the SAW analog signal $x(t)$ launched by a transmitting comb (shown as T in the diagram) to produce a set of small receiver currents i_k in a matrix of detectors. In this 4×4 square array of detectors the signals are effectively carrying out the Hadamard matrix operation H_4 shown by eqn (3.58). The changes in sign are obtained by arranging to collect the output at each of the detectors with a phase of 0 or π rad. This is indicated in the diagram by the plus and minus signs at each transducer position. Thus a one-dimensional forward transform of the input signal $I = H \cdot i$ is formed by line summation to give line output values $I_1 - I_4$. The first term I_1 is obtained by summing the electrical signal when the acoustic waveform launched by T is present under transducers, 1, 2, 3, and 4 at time t_0 to give $i_1 + i_2 + i_3 + i_4$. The second term I_2 gives $-i_5 + i_6 - i_7 + i_8$, and so on. A parallel read-out of the transformed vector I_{1-4} is thus obtained for time t_0. Henaff [61] and Rebourg [62] describe practical

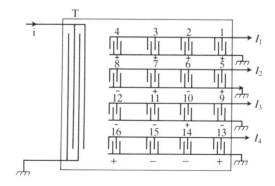

Fig. 6.11 A SAW Hadamard matrix

applications of Hadamard SAW transform devices based on multiple operation of this basic transform.

A two-dimensional method of using SAW devices for image transformation is described by Kornreich and Kowel [63] and implemented by Hannigan [64] for terrain feature classification. This operates in a quite different manner to the transversal filter devices we have been considering. The basic idea behind this interesting application is described with reference to Fig. 6.12. An acoustic wave is propagated along the surface of a film of cadmium suphide (Cd S) forming a substrate onto which an optical image is projected. The Cd S film reacts to the acoustic wave and the light image in two different ways, and it is the interaction of these that effects the transformation.

The propagating acoustic wave is converted into a small current at the detecting filters by a reverse piezo-electric effect. This alternating current is proportional to the amplitude of the signal which launched the acoustic wave along the substrate. The impact of the light image however modifies the value (modulates) these currents through a change in photoconductivity, which is related to the intensity of the light in a given image area. By making the signal a sinusoidal waveform and varying its frequency with time (i.e. creating a chirp waveform) then the amplitude of these small currents will reflect the frequency content of the image on a line-by-line basis as the acoustic wave traverses across the image. This is in contrast to the normal two-dimensional DFT, which calculates the complete image in an unscanned form and does not produce the frequency content until the entire transform has been carried out. Further details of this method may be found in the references already cited.

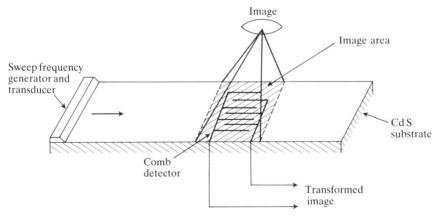

Fig. 6.12 A CCD image processor.

6.5 Programmable signal processors

The complexity and limited application of many of the hardware FFT devices described in Section 6.2 make their construction uneconomic for all but very specialized applications. The falling cost of VLSI devices has now made it practical to produce a general-purpose device, the *programmable digital signal processor* (PSP), which can carry out not only domain transformation but also signal conversion, averaging, filtering, multiplexing, and other signal processing functions on analog or digital data. There is in fact considerable advantage in carrying out digital operations on converted analog data even if the output is to be converted back to analog form again, since much higher accuracy and reliability are obtained. An important application area here is multi-channel voice band communication, where the sampling rate is 8 kHz—well within the capability of the new technology [65]. Recent work on integrated and programmable signal processors has shown that large scale digital circuits are now competitive for applications with signal bandwidths less than about 50 kHz [66].

Many different designs of PSP are described in the literature. Some are simply high-speed microcomputers limited to carry out multiple digital filtering operations; others are used in data logging situations which may or may not contain transformation. Finally, a growing number of designs are described and sometimes actually constructed to carry out a repertoire of signal processing operations. These are special-purpose microcomputers having an instruction set formed to facilitate repetitive inner-product calculations (see eqn (3.6)) and often accompanied by extensive signal processing software.

Earlier versions of these devices were constructed using standard MSI components and resulted in a large component count, high power consumption, and high cost. The availability of high-speed microprocessors, bipolar bit-sliced components, and monolithic array multipliers have now reduced the component chip count for modular designs. A further reduction in size has been seen in recent years with the development of a single integrated circuit processor or a *single-chip PSP* having external memory [67].

6.5.1 Modular architecture

Since the specialized processing modules incorporated in the PSP perform completely primitive operations, the input/output rates are relatively low. The data path flexibility required may not use VLSI too efficiently due to current chip pin limitations. However the considerable flexibility in control and ability to write signal processing software routines in a high-level language more than compensate for these deficiencies.

Architecturally the PSP is generally designed as a number of independently controlled elements, all of which are strongly parallel in operation. These elements will include an arithmetic unit, an address and decoding function, input/output channel control, and a control processor operating from a stored program. Each element will have a working memory to prevent the common system bottleneck that occurs from the use of a single main memory. Extensive use of ROM and RAM devices is made in programming and control of each module. Reprogramming then requires only rewriting of code or exchanging the ROM.

Conventional data processing demands a very flexible data-dependent control sequence. It is necessary to allow for a wide variety of data routes with many jumps from one stored program area to another and even more complex 'nested' jumps. Signal processing operation control on the other hand, is highly structured, containing few data-dependent jump sequences. Hence fixed control programs, held in ROM or PROM, are generally used.

A variety of designs are possible. Several use a number of interconnected microprocessors in which one processor provides the main memory control, a second provides system control, and a third functions as an arithmetic unit [68]. Another division of functions is to differentiate between inner-product and more general operations and to carry these out in separate microprocessor units [69].

Figure 6.13 shows a typical PSP modular design. Here we can identify

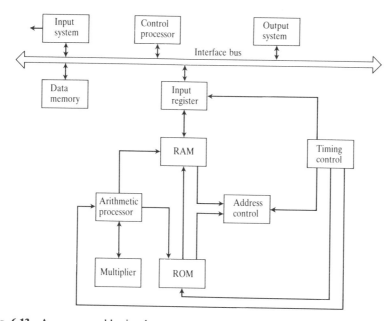

Fig. 6.13 A programmable signal processor.

five major functional elements connected by a main system bus:

    (a) control processor;

    (b) arithmetic processor;

    (c) data memory;

    (d) timing control;

    (e) systems.

The arithmetic processor may well be designed as an inner-product processor, where speed and accuracy are essential. For many operations only a bit-slice implementation is practical with current implementations; this is achieved through a pipeline process as described earlier. A 'look-ahead' operation fetching down several instructions ahead of the instruction currently being executed can also be implemented so that decoding of the earlier instructions can be partially carried out prior to their application [70]. The timing sequencer is a key module in such a design and a detailed overview of its function in one particular PSP is given in [71].

6.5.2 Microprogramming the PSP

Many designs of PSP use a block-structured signal processing language. This is problem-orientated in its architecture and the instruction sets are biased towards signal processing requirements. Thus the block structure permits programs to describe actions by 'statements' and data by 'declaration'. The statement vocabulary will include assignment, case, for, conditional (if, then, else), restricted go-to, return, exit, procedure, and multiprogramming statements. These last include task-manipulation, signal, and synchronization statements and all will include composite statements for carrying out the butterfly and recursive filtering operations, both of which rely on composite multiply-add operations.

No matter what processing capabilities the LSI or single-chip PSP has, there will always be applications that must use more than one of them, so multiprocessor software links will be necessary. Also in some cases the PSP is linked to other larger general-purpose computer systems. To do this the programming facilities include communication software and an operating system compatible with the requirements for signal processing and which could for example, include graphics software. A description of a sophisticated system of this kind is given in [71].

The 'single-chip' PSPs discussed in the next section contain necessarily limited software and instruction sets. Assembly language is used and support software is available to create a virtual machine with which the user can work [72]. Indeed it is usually essential to carry out all the program development and testing with a computer linked to the PSP. Examples of this are a macro-assembler to translate the assembler language into executable object-code, error-diagnostic software, in-circuit simulation to allow verification of the program design before translation, and a compiler, which can permit program development to be carried out in a modular fashion. This latter allows a high-level language (e.g. PASCAL, C, or FORTH) to be used for applications design, to which a number of special function commands or macros may be added [73].

A feature of software support for some PSPs, and more particularly the single-chip PSP, is the availability of custom-designed hardware development systems to assist in the use and sometimes the actual design of the PSP (e.g. 'burning-in' the PROMs or PLAs used in the device). These development systems are microprocessor-based systems that can load programs from a host computer into the chip and will contain editing display facilities to monitor the user-specialized programs [70]. A program simulator may be contained in this or in a second hardware device to ensure that the program functions correctly in an on-line situation before the PSP program is accepted or the PROM within the chip modified to contain the final user program.

6.5.3 The 'single-chip' PSP

One of the earliest attempts to construct a single-chip signal processor was the Bell Labs DSP (digital signal processor) developed in the late 1970s as a telecommunications signalling device [74]. Its principle use is for digital filtering, detection, and modulation of signals. The main emphasis in its design lies in rapid calculation of large numbers of multiply/add operations performed by the parallel pipeline computer architecture. The organization of the DSP consists of a data arithmetic unit, an address arithmetic unit and an input/output unit. Auxiliary memory is essential for program development although, once pro-grammed for a given purpose, the DSP functions in a stand-alone fashion. Speed of operation is indicated by the instruction cycle time of 800 ns, enabling about a million high-precision arithmetic (36-bit product) computations per second.

A similar device, the Intel 2920, functions mainly as a multiple digital filter but can also carry out spectral analysis through heterodyning a signal through a narrow bandpass filter [75]. The configuration of this single-chip PSP contains built-in analog–digital and digital–analog con-verters, enabling it to function as a high-resolution analog signal processor. The EPROM program storage of the 2920, like the Bell DSP, is small and an auxiliary support microprocessor is required to develop and load working programs.

A characteristic of the Intel 2920 and later PSPs is that the program to be executed is held quite separately from the data on which it operates, enabling a complete overlap of instruction fetch and execution to be obtained. This has become known as a *Harvard architecture* [76]. These single-chip PSPs were designed to replace complex analog signal processing operations, which demanded multiple high-precision fitering.

More recently single-chip PSPs have become available designed for wider application to include fast Fourier transformation. One of the first of these was the NEC 7720 signal processing interface, which incorporates a considerable amount of parallel processing, making use of a pipeline architecture [77]. The processor is controlled by directly executable 23- bit microinstructions where the fetching and decoding of the next instruction is carried out concurrently with the execution of the present instruction. The clock rate is 8 MHz. Using separate memories for instructions, fixed data, (filter coefficients, FFT twiddle factors and angular coefficients etc.), dynamic data (in RAM), and scratch-pad memory enables full use to be made of the parallel operation for this device.

A major application for the NEC 7720 lies in multiple digital filter operation, and a number of detailed designs for IIR and FIR filters are described by Zoicas [78]. It is also effective as a fast FFT device carrying out a 64 complex value transformation on a single chip with arrangements for increasing the transformation word to 1024 points by device interconnection. The total processing time for 1024 points, including a bit-reversal routine, is about 77 ms.

The Texas Instruments TMS 320 single-chip PSP provides a variety of options for system configuration to carry out a FFT, providing a trade-off between memory and speed for a given word size. It is particularly valuable for spectral analysis and finds wide use in spectral analysers, speech analysis, pattern recognition, and sonar processing [70]. A single 64 complex-value FFT may be executed on a single chip in 580 μs. In a multiprocessing configuration a word length of 1024 points is transformed in 33 ms by using off-chip RAM to store the intermediate 1024 values.

A functional schematic diagram for the TMS320 is shown in Fig. 6.14 (by courtesy of Texas Instruments Inc). A fast arithmetic and logic unit operates with 16/32-bit logic and the parallel hardware multiplication performs 16×16 bit two's complement multiplication in 200 nS. It is this very fast multiplication rate that supports FFT computation at speeds that are acceptable for on-line processing. Referring to Fig. 6.14 the data distribution is as follows: the 16-bit T register stores temporally the multiplicand and the P register stores the 32-bit result. Multiplier values are obtained from the data memory or are derived directly from the multiplier instruction word. This multiplier also allows other operations such as convolution, correlation and filtering at rates up to 2.5 M samples s^{-1}.

Due to their limited program storage capability single-chip PSPs usually need access to eternal memory for a given application and in the case of program development, access to an external microprocessor.

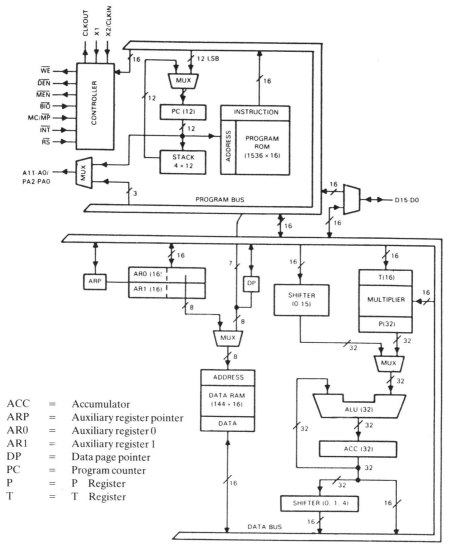

Fig. 6.14 The Texas Instruments digital signal processor. (by courtesy of Texas Instruments Ltd.)

ACC	=	Accumulator
ARP	=	Auxiliary register pointer
AR0	=	Auxiliary register 0
AR1	=	Auxiliary register 1
DP	=	Data page pointer
PC	=	Program counter
P	=	P Register
T	=	T Register

Consequently the speed of external data transfer becomes important, particularly for the external memory required in an on-line situation. The TMS320 enables access time for up to 8 kb of off-chip memory to be the same as would occur for access to its internal memory, namely at an instruction rate of 5 Mb s^{-1}. An additional high-speed burst rate of 40 Mb s^{-1} is also obtained through the 16-bit parallel data bus.

The TMS320 is programmed in assembler code for its 60 member instruction set. As with other PSPs support hardware is necessary to program and test the device. This consists of a macro assembler/linker to convert the assembly language into object code and to insert user macroinstructions; a simulator and an on-line emulator which provides linkage to a host computer or peripherals (e.g. terminals, printer etc.).

Single-chip PSPs are presently limited by the size of internal memory, use of assembler language, and the need to use ancillary equipment for program development, rather than speed of operation. It is to be expected that continual development in VLSI chip density and architecture will eventually enable a more self-contained special-purpose processor to be made available which will greatly reduce the present cost and widen further the applicability of signal processing methods.

References

1. Chu, S. A prime factor algorithm using distributed arithmetic. *IEEE Acoustics Speech Signal Process.* **ASSP-30** (4), 217–27 (1982).
2. Figueiredo, A. C. D., Sa Marta, E. and e Selver, J. G. C. Fast transposition technique for microprocessor array processing attachments. *Euromicro Symposium Proceedings.* European Research Office, Bell Telephone, Belgium. pp. 25–30 (1980).
3. Dimond, K. R. Bit-slice processing. In *Digital signal processing* (ed. N. B. Jones). pp. 419–433, Peter Peregrinus, London (1982).
4. Kolb, H. J. A signal processor using bit-slice elements for the audio-frequency range. *Signal Process.* **2**, 339–46 (1980).
5. Allen, P. J. and Holt, A. G. J. Implementation of a second-order digital filter using AM2900 bit-slice devices and a fast multiplier. *IEE Proc.* **128G** (4), 216–19 (1981).
6. Gerhauser, H. A fast bit-slice computer for real-time signal processing. In *Electronics to microelectronics* (ed. Kaiser W. A. and Proebster W. E.) North Holland, Amsterdam (1980).
7. Magar, S. S. and Robinson, D. A. Microprogrammable arithmetic element and its applications to digital signal processing. *IEE Proc.* **127F** (2), 99–106 (1980).
8. Mich, J. R. and New, B. J. Bit-slice devices for signal processing IEEE Cat. No CH1559-4/80/0000–0372 pp. 372–5 (1980).
9. Corinthios, M. J. Architecture of signal processors. *Int. J. Mini Microcomput.* **4** (3), 63–9 (1982).
10. Alexandridis, N. A. Bit-sliced microprocessor architecture. *IEEE Computer* **11**, pp. 56–80 (1978).
11. Schirm, L. Complex FFTs yield to single-board computer. *Electron. Des.* **10**, 27 May, 153–158 (1982).
12. Bergland, G. D. Fast Fourier transform hardware implementation—a survey. *IEEE Trans. Audio Electroacoust.* **AU-17** (2), 109–19 (1969).
13. Yuen, C. K., Beauchamp, K. G., and Robinson, G. P. S. *Microprocessor*

systems and their application to signal processing. Academic Press, London (1982).

14. Beauchamp, K. G. *Applications of Walsh and related functions.* Academic Press, New York (1984).
15. Geadah, Y. A. and Corinthios, M. J. G. Natural, dyadic, and sequency order algorithms and processors for the Walsh–Hadamard transform. *IEEE Trans. Comput.* **C-26,** 435–42 (1977).
16. Muniappan, K. and Kitai, R. Walsh spectrum measurement in natural, dyadic, and sequency ordering. *IEEE Trans. electromagn. Compat.* **EMC-24,** 46–9 (1982).
17. Pease, M. C. An adaption of the FFT for parallel processing. *J. Ass. Comput. Mach.* **15,** 252–64, (1968).
18. Shirata, K. and Nakatsuyama, M. The fast Walsh–Hadamard transform and processors using new permutation networks. *Trans. IECE Japan* **J63-D,** 319–25 (1980).
19. Nakatsuyama, M. and Nishizuka, N. The fast Walsh–Hadamard transform and processors usng delay lines. *Trans. IECE Japan* **E64,** 708–15 (1981).
20. Johnston, J. A. Parallel pipeline FFT. *IEE Proc.* **130F** (6), 564–70 (1983).
21. Groginsky, H. L. and Works, G. A. A pipeline fast Fourier transform. *IEEE Trans. Comput.* **C-19** (11), 1015–1019 (1970).
22. O'Leary, G. A high-speed cascade fast Fourier transformer. IEEE Workshop on Digital Filtering, Arden House, Jan. (1970).
23. Ashouri, M. R. and Constantinides, A. G. A pipe-line fast Walsh–Fourier transform. *IEEE Conference on Acoustics, Speech, and Signal Processing,* Hartford, Conn., pp. 515–18 (1977).
24. Luk, W. K., and Li, H. F. Microcomputer-based real-time/on-line FFT processor, *IEE Proc.* **127E** (1), (1980).
25. Wallingford, E. E. and Collins, W. R. A dynamic electroencephalogram frequency analyser. *IEEE Trans. Instrum. Meas.* **IM-27,** 70–3 (1978).
26. Kobylinski, R. A., Stigall, P. D., and Ziemer, R. E. A microcomputer-based data acquisition system with hardware capability to calculate a fast Fourier transform. *IEEE Trans. Acoust. Speech Signal Process.* **ASSP-27,** 202–3 (1979).
27. Kolb, H. J. A signal processor using bit-slice elements for the audio-frequency range. *Signal Processing,* North Holland, **2,** 339–346 (1980).
28. Stigall, P. D., Ziemer, R. E., and Pham, V. T. A performance study of 16-bit microcomputer implemented FFT algorithms, *IEEE Micro,* **2,** pp. 61–6 (1982).
29. Despain, A. M. Very fast Fourier transform algorithms hardware for implementation. *IEEE Trans. Comput.* **C-28,** 333–41 (1979).
30. Gibson, R. M. and McCabe, D. P. Fourier transform algorithm implementations in a general-purpose microprocessor. IEEE Cat. No. CH1610-5/81/0000–0670, pp. 670–2 (1981).
31. Said, S. M. and Dimond, K. R. Improved implementation of FFT algorithms on a high performance processor. *Electron. Lett.* **20** (8), 347–9 (1984).
32. Durgan, B. K. and Lai, D. C. A microprocessor implementation of the fast Walsh transform. IEEE Symposium on Applications on Mini and Microcomputers, Philadelphia, pp. 395–9 (1980).
33. Collado, F. M., Romero, A. M. B., and Martin, J. A. M. Fast Haar

transform algorithm; a 6809-based real-time evaluation. *Microprocess. Microsyst.* **8** (3), 126–35 (1984).

34. Muniappan, K. and Kitai, R. Microprocessor-based Walsh–Fourier spectral analyser. *IEEE Trans. Instrum. Meas.* **IM-28,** 295–9, (1979).

35. Berauer, G. Fast in-place computation of the discrete Walsh transform in sequency order. Proceedings of a Symposium on Applications of Walsh Functions, Washington, DC, AD744650, pp. 272–5 (1972).

36. Abelson, H. and Andreae, P. Information transfer and area–time trade-offs for VLSI multiplication. *Communs Ass. Comput. Mach.* **23,** 20–23 (1980).

37. Savage, J. Area-time trade-offs for matrix multiplication and related problems in VLSI models, *J. Comput. Syst. Sci.* **22,** 230–242 (1981).

38. Feridun, A. M. Torng, H. C., and Li, H. A methodology for the design of VLSI -based FFT multiprocessors. Proceedings of 17th Hawaii International Conference on Systems Science, Honolulu, Vol. 1, pp. 12–21, Jan. (1984).

39. Vuillemin, J. A combinatorial limit to the computing power of VLSI circuits. *Proceedings of 21st Symposium on foundations of computer science.* IEEE Computer Society, pp. 294–300 (1980).

40. Thompson, C. D. Fourier transforms in VLSI. *IEEE Trans. Comput.* **C-32,** (11), 1047–1057 (1983).

41. Johnsson, L. and Cohen, D. Computation arrays for the discrete Fourier transform. Digest IEEE Compcon, pp. 236–244 (1981).

42. Kung, H. T. Why systolic architecture? *Computer* **15** (1), 37–46 (1982).

43. Thompson, C. D. Generalized connection networks for parallel processor intercommunication., *Proc. 20th Ann. Symp. Found. Comput. Sci.* **C-27,** 1119–25 (1978).

44. Feridum, A. M. A methodology for the design of VLSI-based FFT multiprocessors. Ph.D. thesis, Cornell University, Aug. (1983).

45. Nick, H. Binary logic Walsh function generator. *IBM Tech. Disclosure Bull.* **22,** 4650–1 (1980).

46. Smith, E. G. CAD design of a modularized fast Walsh -Hadamard transformation device. *IEEE Proc. Int. circ. Comput.* 916–20 (1980).

47. Buss, D. D., Tasch, A. F., and Benton, J. B. CCD applications to signal processing. In *Charge-coupled devices and systems* (ed. M. J. Howes and D. V. Morgan) Wiley, New York, pp. 79–88 (1979).

48. Buss, D. D., Veenkant, R. L., Brodersen, R. W., and Hewes, C. R. Comparison between the CCT, CZT, and the digital FFT. In *Charge-coupled devices: technology and application.* (ed. R. Melen and D. Buss). IEEE Press, New York (1976).

49. Brodersen, R. W. and Hewes, D. C. R. A 500-stage CCD transversal filter for spectral analysis. *IEEE Trans. electron. Devices* **ED-23,** 143–52 (1976).

50. Jack, M. A., Park, D. G., and Grant, P. M. CCD spectrum analyser using the prime transform algorithm. *Electron. Lett.* **13** (15), 431–2 (1977).

51. Kapur, N., Maron, J., and Jack, M. A. Discrete cosine transform processor using a CCD programmable transversal filter. *Electron. Lett.* **16** (4), 139–41 (1980).

52. Roberts, J. B. G., Darlington E. H., Edwards, R. D., and Simons R. F. Transform coding using charge-coupled devices, *Electron. Lett.* **13** (10), 277–8 (1977).

53. Yarlagadda, R. and Hershey, J. E. Architecture of the fast Walsh–

Hadamard and fast Fourier transforms with charge transfer devices. *Int. J. Electron.* **51,** 669–81 (1981).

54. Jack, M. A. The theory, design and applications of surface acoustic-wave Fourier transform processing; a review. *IEEE Proc.* **68** (4), 450–68 (1980).

55. Alsup, J. M. Surface acoustic wave CZT processors. *Proceedings of the IEEE ultrasonics symposium* pp. 378–81. IEEE Publication No. 74 CHO 089–1SU (1974).

56. Atzeni, C., Manes, G., and Masotti, L. Programmable signal processing by analog chirp transformation using SAW devices. *Proceedings of the IEEE ultrasonics symposium* pp. 371–6. IEEE Publication No. 75 CHO 994–4SU (1975).

57. Otto, O. W. Real time Fourier transform with a surface wave convolver. *Electron. Lett.* **8** (25), 623–5 (1972).

58. Alsup, J. M. Prime transform SAW device. *Proceedings of the IEEE* pp. 377–80 IEEE Publication No. 75 CHO 994-4SU (1975).

59. Klauder, J. R. The theory and design of chirp radars. *Bell Syst. Tech. J.* **39** (4), 745–808 (1960).

60. Collins, J. D. Sciarretta, W. A., and Macfall, D. S. Signal processing with CCD and SAW technologies. *Proceedings of the IEEE ultrasonics symposium* pp. 441–50. IEEE Publication No. 76 CH1 120 SSU (1976).

61. Henaff, J. Image processing using acoustic surface waves. *Electron. Lett.* **9,** 102–4 (1973).

62. Rebourg, J. C. New Hadamard transformer. *IEEE Trans. Sonics Ultrason.* **SU-25,** 252–6 (1978).

63. Kornreich, P. G. and Kowel, S. T. DEFT: direct electronic Fourier transform of optical images. *IEEE Proc.* **62,** 1072–87 (1974).

64. Hannigan, J. F. Direct electronic Fourier transform (DEFT) spectra for terrain feature classification. *Soc. Photo. opt. instrum. Engng* **241,** 122–8 (1980).

65. Murano, K. Mochida, Y., Amano, F., and Kinoshita, T. Multiprocessor architecture for voice-band data processing. Int. Conf. Commun. Rec. pp. 37.3.1–37.3.5 (1979).

66. Thompson, J. S. and Tewksbury, S. K. LSI signal processor architecture for telecommunication applications, *IEEE Trans. Acoust. Speech signal Process.* **ASSP-30** (4), 613–31 (1982).

67. Quarmby, D. J., ed. *Signal processor chips.* Granada, London, (1984).

68. Mintzer, L. A microprogrammable signal processor. *IEEE Trans. Acoustics Speech signal Process.* **ASSP-24,** 494–7 (1977).

69. Glass, J. M. A programmable signal processing architecture. *IEEE Trans. Acoust. Speech signal Process.* **ASSP-27,** 702–5 (1979).

70. Boddie, J. R. Overview: digital signal processor, support facilities and applications. *Bell Syst. tech. J.* **60** (7, 2), 1431–9 (1981).

71. Zeman, J. and Troy, N. H. A high-speed microprogrammable digital signal processor employing distributed arithmetic. *IEEE Trans. Comput.* **C-29** (2), 134–44 (1980).

72. So, J. TMS320—a step forward in digital signal processing. *Microprocess. Microsyst.* **7,** 10 (1983).

73. *The 2920 analog signal processing handbook,* Intel, Santa Clara, CA (1980).

74. Special issue: The digital signal processor. *Bell Syst. tech. J.* **60** (Sept.) (1981).

75. Townsend, M. Hoff, M. E., and Holm, R. An NMOS microprocessor for analog signal processing. *IEEE J. solid state Circuits* **SC-15,** 33–7 (1980).
76. Cragon, H. The elements of single-chip microcomputer architecture. *Computer* **13** (10), 27–41 (1980).
77. Nishitani, T., Kawakani, Y., Maruta, R., and Sawai, A. LSI signal processor development for communications equipment. *Proceedings of ICASSP-80,* Vol. 3, pp. 386–9. IEEE Press (1980).
78. Zoicas, A. The NEC 7720. In *Signal processor chips* (ed. D. J. Quarmby) pp. 86–125. Granada, London (1984).

7

SPEECH PROCESSING AND COMMUNICATIONS

7.1 Introduction

Much of the theoretical background of signal processing methods arose during research into speech characteristics and speech communication. It is thus one of the earliest applications of signal analysis using continuous analog methods. Digital signal processing was first applied to speech as a simulation of complex analog systems. With the development of fast computers and digital methods newer techniques have been developed, which in many cases have no realizable counterpart in analog implementation [1, 2].

In recent years the economics in large-scale integration for digital logic have exerted a considerable influence on methods of speech processing and communication [3, 4]. After a period in which considerable ingenuity has been shown in developing transmission methods involving continuous waveforms, interest is now focused on the transmission of digitally encoded information, whether this be speech, vision, or computer-processed data. This is of course, due to the dramatic reduction in the cost of digital hardware and availability of VLSI chip technology described in the previous chapter. Nowhere is this seen more clearly than in speech processing, where improvements in performance need not now be accompanied by considerable increase in cost, despite the high level of complexity that may be involved. This has led to a renewed interest in novel and more sophisticated digital methods that are now realizable in practical application.

The improved processing capability has been matched by new methods of digital communication, thus ensuring fast and reliable transmission of information with all the benefits of digital data handling, not least being the ability to define with extreme precision the accuracy of the information transmitted. The major developments in digital communication have been in multiple signal transmission along a single communication channel and in coding for transmission. This latter is carried out not only to minimize transmission error but also to achieve economy in the use of available signal bandwidth.

Methods of coding can be highly complex and are not particularly illustrative of transform methods [5]. Multiple signal transmission, on the other hand, can benefit considerably from frequency or sequency domain transformation and a number of currently used techniques will be considered later in this chapter.

7.2 Speech processing

The two major areas of speech processing are those of speech coding for data compression and speech recognition through signal analysis. Other areas of interest are found in speech enhancement [6] and in voice response devices [7, 8]. The subject has grown enormously in recent years and it will only be possible to present an outline of the work in this chapter, with the emphasis placed on transform methods for speech compression and recognition. Both involve analysis of human speech using time/frequency transformation and use a generalized model for the mechanism of speech production. The model is based on the manner in which we actually utter comprehensive sounds, and in using the model it is often possible to discern some simplified ways of reproducing such sounds using electronic techniques.

7.2.1 A model for speech production

A basic assumption of most physical models for speech production is that the sources of excitation and production of sounds are linearly separable. Thus in the simplified model of Fig. 7.1 the sound source (lungs and larynx) is assumed to produce two forms of excitation; a quasi-periodic sequence of vibrations giving rise to the 'voiced' speech sounds (such as vowels) and a random sound produced by a turbulent air flow (equivalent to white noise) resulting in 'unvoiced' speech sounds [9]. These latter can take the form of fricative consonants, typified by the sound 's', responsible for the 'hissing' quality in speech; and the stop consonants

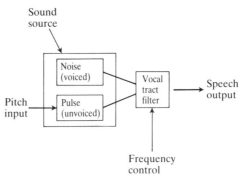

Fig. 7.1   A model for speech production.

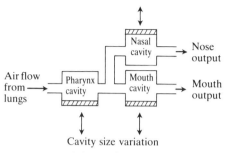

Fig. 7.2 Simplified representation of the vocal tract.

obtained by building up pressure behind a closure such as the lips and suddenly releasing it. Most of the sounds of speech can be modelled quite adequately by the response of three loosely coupled resonant cavities to either a controlled periodic series of impulses or to a noise source (Fig. 7.2). The physical configuration of the vocal tract itself is highly variable and its dimensions depend on the position of the articulators (tongue, larynx, etc.). In general terms the vocal tract consists of a non-uniform tube about 17 cm in length, giving its first quarter wavelength resonance at about 500 Hz. This represents the running *pitch* frequency of speech. The content of speech arises from relatively slow variations in the resonant frequencies of the cavities shown in Fig. 7.2, which result in certain normal modes of resonance for the voiced speech signals, known as *formants* [9, 10].

7.2.2 Speech coding

Digital encoding of continuous (analog), or indeed any other variable amplitude signal can be carried out through a variety of digital codes, some of which can be optimized for greater coding efficiency. This is typically achieved by making use of observed statistics of a given group of signals, such as speech waveforms, and referred to as *waveform coding*. A second class of speech coders are those where a *description* of the signal is transmitted rather than the signal itself. Thus we might use the characteristics of the speech model described in Section 7.2.1 or alternatively transmit the spectral shapes of a succession of speech sounds or *phonemes* that constitute speech. Phonemes are the basic linguistic elements taken from a finite set of distinguishable mutually exclusive sounds. They can be considered as a coded set of speech symbols, which we use in ordinary speech. All these various forms of encoders are called *source coders* and, when applied to speech, are known as voice coders or *vocoders*.

7.2.3 Waveform coders

Waveform coders can operate in the time or the frequency domain. Examples of the former are the well-known pulse code modulation (PCM) or the differential pulse code modulation (DPCM), which exploits the high degree of correlation existing between adjacent samples of a speech signal by coding the *difference* between waveform amplitudes and integrating the quantized difference samples [11]. A special case of DPCM is a one-bit version known as *Delta modulation,* in which only the polarity of this difference is noted and transmitted [12]. These coding algorithms apply to the speech waveform in its time domain representation as a sampled and quantized analog signal. Another class of waveform encoding algorithm operates in the frequency domain by dividing the speech signal into a number of separate frequency components and encoding these components separately. This is known as the phase or *sub-band coder* (SBC). Here the speech band is divided into a small number of sub-bands by means of a set of bandpass filters or through short-term spectral analysis and transformed down to zero frequency by a process equivalent to single side-band modulation [13]. It is then possible to sample each band separately at its minimum Nyquist rate (i.e. at twice the width of the sub-band) and to transmit the multiplexed sets of sampled values, still in the time domain (Fig. 7.3). Due to the modulation carried out previously, the sampling rate will effectively vary for the real signal frequencies, being kept at roughly the Nyquist level for the entire bandwidth of the signal. This reduces the number of bits transmitted across the systems and has other advantages in terms of reduction in quantization and other additive noise [14].

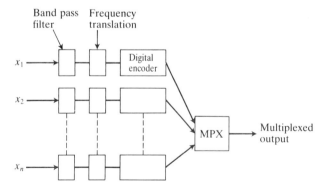

Fig. 7.3 A sub-band coder.

7.2.4 Adaptive transform coding

One of the characteristics of speech is that it produces a frequency spectrum that does not remain constant with time. A method of sub-band coding that takes this into account is known as *adaptive transform coding* (ATC). The sampled speech signal is divided into a series of short time series segments, each of which is transformed separately into the frequency domain before quantization, thresholding, and transmission. At the receiving end the quantized coefficients are inversely transformed and assembled into segment order to reproduce the original signal. Not only are fewer transform coefficients required to reconstitute the original signal compared with the number of coefficients required in a time series representation for the same error, but the technique allows for variation in the numbers of bits allocated in the quantization process to different frequency coefficients. One such variation is to take into account the weighting or importance of the frequency spectrum in the formant characteristic of speech. Figure 7.4 shows a typical smoothed frequency response of normal speech. The formant frequencies are shown as F_1, F_2, and F_3, and correspond to the resonant frequencies of the acoustic cavities shown in Fig. 7.2. The number of bits used to code or quantize the frequency coefficients is made proportional to the magnitude of this shape. This adjusts the bit allocation in accordance with the intensity value of the speech sequence, improving the representation and matching the actual mechanism of speech production.

The discrete cosine transform (DCT) described in Section 5.6.1 has been found to be well-suited for speech coding due to its mean-square transformation efficiency and even symmetry. An ATC system is described by Zelinski and Noll that makes use of a fast cosine transform together with an adaptive bit assignment and adaptive quantization [15]. Comparisons are given for a number of different unitary transformations that may be used in ATC. These are compared with conventional PCM

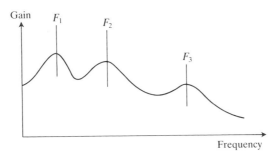

Fig. 7.4 Frequency weighting of the speech model.

Table 7.1   SNR gains for various unitary transforms

Gain(dB)	Block size = N			
	16	32	64	128
KLT	8.1	8.9	9.1	9.6
DCT	7.8	8.5	9.0	9.1
DFT	5.8	6.8	7.6	9.2
WHT	3.1	3.1	3.2	3.2

waveform encoding in terms of the SNR. It is shown that while the KLT may be considered as optimum, the DCT gives a very similar performance and, of course, may be implemented by means of a fast transformation. The DCT is also shown to give better SNRs than either the DFT, DST, or the WHT. This is in agreement with the general results given earlier by Pratt and discussed in Section 2.5.1. The SNR gains for the various unitary transforms are shown in Table 7.1 (taken from Zelinski and Noll's results) for segment block sizes of $N = 16$, 32, 64, and 128.

7.2.5   Vocoders

As stated earlier the vocoder transmits values for a set of the characteristics of a speech signal rather than samples taken directly from the signal itself. The method therefore depends on finding a suitable model for speech that is sufficiently flexible to take into account the full range of the human voice irrespective of age, sex, race or the speech environment (which may be noisy). Such models as are available fall short of this ideal, a major reason being the absence of adequate emphasis or linguistic stress. The emphasis in speech is important, since it reflects the meaning of the words uttered. It is not easy to incorporate this into a speech model, so vocoder speech is often mechanical in quality even if highly intelligible [16]. However the increase in efficiency for vocoder transmission can be very great, thus justifying its implementation in many cases. Whereas for PCM coding a sampling transmission rate of about 64 kb s^{-1} is needed for intelligible speech, this may be reduced to about 1 kb s^{-1} when only the source model parameters are transmitted [17]. Hence considerable interest has been shown over the last decade in developing usable vocoder methods and in improving their linguistic quality. An early method was the *channel vocoder*, which exploits the insensitivity of the hearing mechanism to phase, so that only the envelope of the signal is selected for transmission [18, 19]. A simplified diagram for

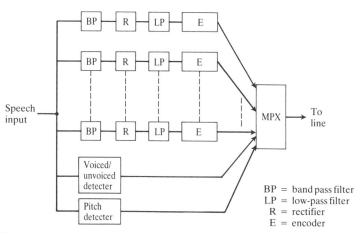

Fig. 7.5 A channel vocoder.

this method is shown in Fig. 7.5. The speech signal is separated by a number of bandpass filters into several contiguous frequency bands as in the SBC coder. The output of each filter is subject to detection and smoothing, thus extracting the envelope of the modulated segment from the signal. This is sampled and multiplexed with other detected segments for transmission. In addition to the contiguous speech channels, the vocoder transmits certain components derived from signal analysis to indicate the mode of excitation, i.e. voiced or unvoiced, together with the average pitch value of the segment. This is necessary to give speech its characteristic tone variation, without which the synthesized speech would be monotonous and lifeless [20]. The total information bandwidth transmitted is approximately 300 Hz, about 1/10 of the original signal bandwidth. At the receiving end the pitch and voice/unvoiced levels are extracted and used to control the output from a pulse and a noise generator, respectively. These noise outputs are fed through a set of bandpass filters and caused to modulate the signal output derived from each of the detected segment characteristics. Thus the channel decoder follows fairly closely the model for speech production shown in Fig. 7.1. An improved channel detector is obtained by specifying only the frequencies of the formants and their amplitudes as a function of time, thus following a *linguistic model* of speech production. These are known as *formant vocoders,* considered later in connection with speech synthesis.

7.2.6 Pitch-recognition vocoders

A difficulty with the channel vocoder and other methods involving the separation and transmission of the pitch of a signal lies in recognition of the value of the pitch of the signal itself, particularly the beginning and ending of a period of relatively constant pitch, the *pitch period*. Given this information, the spectrum within each pitch period can fairly easily be identified and transmitted along with the pitch information for accurate synthesis of the original signal at the receiver. A considerable literature is available on methods of pitch detection, and a comprehensive treatment of a number of these is given in [21, 22]. One of the most successful of these is the use of the cepstrum introduced very briefly in Chapter 1. The method is in essence very simple and depends on an observation made by Noll, who showed that for voiced speech input the cepstrum shows a peak at the fundamental period (pitch) of the input speech signal. No such peak appears in the cepstrum for an unvoiced speech segment [23]. In a practical application the cepstrum is computed at about every 10 to 20 ms and searched for a peak in the vicinity of the expected pitch period (about 1/500 s, as mentioned earlier). If the peak found is greater than a given threshold value, the input speech segment is likely to be voiced and the position of the peak will indicate the period. Absence of a peak indicates the likelihood that that particular segment of the speech signal is unvoiced.

7.2.7 Linear prediction

A powerful method of signal analysis which has led to an effective vocoder design is that of *linear predictive coding* (LPC) [24]. This method is also notable in determining the impulse excitation points in voiced speech and hence the pitch period, described previously as the key to vocoder signal synthesis [25]. Essentially the idea behind LPC is to approximate a speech sample through a linear combination of weighted past speech samples. The sum of the squared differences between this approximate value and the actual speech sample is taken over a finite interval (segment) and minimized to enable a unique set of predictor weights to be determined. In the LPC, vocoder speech is synthesized using a vocal tract model by estimation of the basic speech parameters; pitch, formants, spectra, and the physical characteristics of the vocal tract itself. The end product is a representation of speech for transmission and storage having a very much smaller digital representation than may be obtained by straightforward digitization of the original speech. A simplified speech model for this method is shown in Fig. 7.6. The spectral characteristics of the complex vocal tract (Fig. 7.2) are here represented by a time-varying digital filter whose coefficients, a_k, need to be determined. Other parameters needed for the model are the voiced/unvoiced classification (position of the switch shown in Fig. 7.6), the pitch period, and a gain parameter, G. Since all these parameters vary slowly with time, a segment by segment analysis/synthesis of the speech

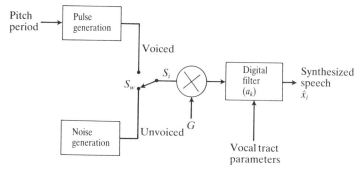

Fig. 7.6 A linear predictor vocoder.

signal is made. The pitch period and voiced/unvoiced classification are estimated by one of a number of methods discussed earlier [1]. For the system of Fig. 7.6 the actual speech samples x_i are expressed in terms of the excitation input (pulse or noise) s_i by the simple difference equation

$$x_i = \sum_{k=1}^{p} a_k x_{i-k} + G s_i. \tag{7.1}$$

This will be recognized as containing an IIR filter representation having filter coefficients a_k. Estimation of these coefficients forms the basis of the method of linear predictor analysis.

The predictor error for a pth order system is defined as

$$e_i = x_i - \sum_{k=1}^{p} a_k x_{i-k}. \tag{7.2}$$

If the speech signal corresponds exactly to the model of eqn (7.1), then

$$e_i = G s_i. \tag{7.3}$$

The basic approach in the estimation of the a_k coefficients is to minimize the short-time average prediction error over each speech segment through the solution of a set of linear equations. This error is defined as

$$E_i = \sum_m e_i^2(m) \tag{7.4}$$

$$= \sum_m \left[x_i(m) - \sum_{k=1}^{p} a_k x_i(m-k) \right]^2. \tag{7.5}$$

Here $x_i(m)$ refers to a segment m of the speech signal.

A set of values for a_k that minimizes E_i is found by setting $\partial E_i / \partial a_r = 0$ for $r = 1, 2, \ldots, p$ to obtain a set of simultaneous linear equations

$$\sum_m x_i(m-r) x_i(m) = \sum_{k=1}^{p} \hat{a}_k \sum_m x_i(m-r) x_i(m-k) \tag{7.6}$$

where $\hat{a}_k$ are the required estimated values. Solution of these equations to determine a_k may be obtained from the summated products of present and past signal segments. Comparison of the left-hand side of eqn (7.6) with eqn (1.25) indicates a similarity that is exploited in one method of determining a_k. In this method a set of auto-correlation functions is determined related to pairs of x_i values, i.e.

$$E_i = R_i(0) - \sum_{k=1}^{p} a_k R_i(k) \tag{7.7}$$

where $R_i(k)$ is the short-term auto-correlation between two segments of the signal. This results in a matrix of values for R_i over a range of p segments, which can be solved to give a set of filter coefficients a_k [26]. Finally, an approximate value for the gain G may be derived from eqn (7.3) as the ratio of the error signal and the excitation value, i.e. e_i/s_i. However since this relationship implies a perfect match between the original and synthesized speech value it is not a reliable estimate. Instead the energy in the error signal is equated with the energy in the excitation input and this, together with the assumption that s_i is a zero-mean, white-noise process, enables a more accurate value of G to be obtained. The predictor error e_i itself can be used as one method of pitch detection, since this will be large (for voiced speech) at the beginning of each pitch period. Hence it is only necessary to detect the position of the values of e_i, which exceed a given threshold value, and to note the period between such pairs of values to determine the pitch period.

Alternative methods for the derivation a_k and the other speech parameters are the use of covariance between segments, inverse filter formulation, a lattice filter method, or a maximum likelihood solution. It is not possible to consider these here, but the reader will find a detailed discussion in [1, 24, 27].

7.2.8 The homomorphic vocoder

The term *homomorphic processing* is applied to a class of systems that obey the general principle of superposition. The importance of this type of processing is that an operation, e.g. Q in Fig. 7.7(a), can be decomposed into a cascade of simpler operations, e.g. A, L, and A^{-1} in Fig. 7.7(b). Here A and A^{-1} are direct and inverse transform systems (e.g. Fourier transforms) and L is a linear time-invariant system such as a filter. The cepstrum is one example of homomorphic processing, which we have seen may be used for pitch detection. This is extended in the *homomorphic vocoder*, where the cepstrum is used to separate the excitation information (voiced and unvoiced), pitch, and formant spectral information into the high and low time regions of the cepstrum analysis [28].

(a)

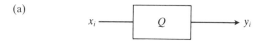

(b)

Fig. 7.7 Homomorphic processing.

To understand why the cepstrum is used for this we need to appreciate that these various components of a speech signal are not just *added* together to form the composite signal, but are *convolved* together. In other words, for two signals f_a and f_b the convolved signal is

$$f_c(t) = f_a(t) \star f_b(t). \qquad (7.8)$$

If this is expressed in the frequency domain in terms of the power spectrum, we might have

$$|F_c(\omega)|^2 = |F_a(\omega)|^2 \, |F_b(\omega)|^2 \qquad (7.9)$$

where the frequency contributions of $f_a(\omega)$ and $f_b(\omega)$ are not separable from each other directly unless they happen to occupy separate regions in the frequency spectrum. If however the logarithm of the power spectrum is examined the two spectra become additive, i.e.

$$\log_e |F_c(\omega)| = \log_e |F_a(\omega)| + \log_e |F_b(\omega)|. \qquad (7.10)$$

It then becomes possible to separate them in the inverse-log-frequency (i.e. pseudo-time) domain through the equivalent of 'frequency' filtering, which will be called *time filtering* in the cepstral domain. In Section 1.5.5 the cepstrum was defined as the Fourier transform of the logarithm of the *power spectrum* of the signal. This was acceptable in initial applications of the cepstrum, as described by Bogert, Healey, and Tukey [29], since these were directed towards the detection of echoes rather than convolution. As a consequence the retention of phase information was not important and the logarithm of real positive values could be used. In cepstral homomorphic processing for deconvolution, the transform system A is such that if the input is a convolution of two signals x_1 and x_2,

$$x_i = x_1 \star x_2 \qquad (7.11)$$

then the z-transform of the input is the product of the corresponding z-transforms for x_1 and x_2

$$X(z) = X_1(z)X_2(2). \qquad (7.12)$$

Since the system is linear we would expect the z-transform of the output to be an additive combination of z-transforms. This property is expressed in eqn 7.13 as

$$
\begin{array}{c}
x_i = x_{i1} \quad \bigstar \quad x_{i2} \\
\big| \qquad \big| \\
z\text{-trans.} \quad z\text{-trans.} \\
\downarrow \qquad \downarrow \\
X_{i1}(z) \cdot X_{i2}(z) = X_i(z) \\
\text{inverse } z\text{-trans.}
\end{array}
\tag{7.13}
$$

and follows the relationship already stated by eqn (1.13) in Chapter 1. The representation of a homomorphic system as described by eqn (7.10) implies that the logarithm must be defined so that it has the property that the logarithm of a product is equal to the sum of the logarithms. To ensure this is so for the complex logarithm described by the z-transform we need to state an inverse transform of a complex logarithm of the Fourier transform

$$
K_c = \frac{1}{N} \sum_{k=0}^{N-1} (\log_e X_k) W^{-kn}
\tag{7.14}
$$

where X_k is the DFT of x_i and $W = \exp(-j2\pi/N)$. The complex cepstrum differs from the power cepstrum defined by eqn (1.36) since it requires the use of a complex logarithm to evaluate the spectral magnitude and phase information. However if the input speech itself is considered as a real sequence having minimum phase, then the complex cepstrum K_c is also a real sequence and can be calculated from the inverse Fourier transform of the logarithm of the *Fourier transform* of the signal. The complex cepstrum is computed by calculating the DFT, its logarithm, and an IDFT at a rate of once every 10 to 20 mS. The pitch and voice information is extracted from the high-time part of the cepstrum as described earlier. The low-time part of each cepstrum is selected by a window w carrying out the time filtering. The resulting speech data is quantized for transmission. A reciprocal arrangement is used for the synthesizer, in which an exponential is substituted for the logarithmic operation and the result convolved with an excitation generator responding to the pitch and voice information transmitted separately. A diagram showing the complete analyser/synthesizer homomorphic vocoder is shown in Fig. 7.8.

A limitation with the use of the cepstrum in implementation is that the phase of the Fourier transformation is lost so that exact synthesis of the speech waveform cannot be obtained. The impulse response at the output

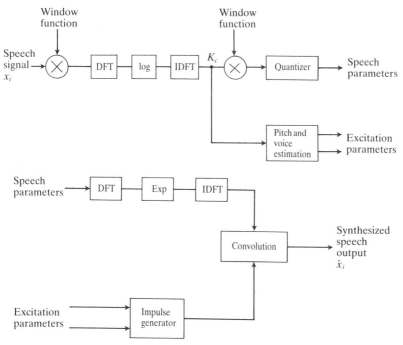

Fig. 7.8 A homomorphic vocoder. (a) Analyser; (b) synthesizer.

of the quantizer will however be a minimum phase synthesis, which is a close approach to natural speech, so this is not a serious limitation. Very low information rates are achieved with the homomorphic vocoder at the expense of some complexity in its hardware realization. This latter is becoming of less importance now that the transform operations may be carried out through the use of a compact VLSI chip.

7.2.9 Speech synthesis

The preceding sections have been concerned with complete vocoder analyser/synthesizer systems. In recent years there has been considerable activity in developing compact synthesizer hardware that can be used for such applications as automatic telephone messages, spoken warning messages in automobile control, lift operation, and other machinery control. Speech synthesizers are also used in 'speak and spell' computers for educational purposes, video games, and as an adjunct to future generations of voice-operated computers now under development. The object of speech synthesis is of course the possibility of storing a vocabulary of words and phrases in a considerably smaller storage space than would be required, for example, in simple waveform coding.

Compact coding methods are necessary in these applications to exploit the redundancy in human speech and to reduce the complexity and cost of speech reproduction.

Single-chip speech synthesizers are now becoming available from several manufacturers. The majority of present commercial chips available use LPC synthesizer chips and many apply the Texas Instruments TMS5000 series in a compact device fabrication. General-purpose VLSI chips discussed earlier in Chapter 4, such as the AMD2901, NEC7720, and the TMS320, have also been applied. Derivation of the time-varying filter coefficients for LPC operation are made through a lattice filter [27], which requires a rather smaller logic configuration than the correlation/convolution methods described earlier.

Earlier designs made use of a number of interconnected standard signal processor chips. A compact design using four AMD2900 chips augmented with a four-channel multiplexer is described by Hofstetter et al. [30]. Problems associated with the implementation of the LPC using standard signal processing chips are the limited input–output capability and limited program memory. A discussion of these problems and techniques for applying the general-purpose chips in LPC analysis are given in [31].

Although allowing compact coding application the LPC synthesis gives rather poor quality for the reproduced speech in its single chip implementation. It is also rather difficult to include a wide vocabulary of speech utterances with the limited storage possible in this method. A more realistic quality of speech and possibility of synthesis for a wide variety of speech phrases is obtained through a linguistic description of the speech [32]. Such synthesis is known as *formant synthesis*. Here the vocal tract is considered as a series of acoustic cavities, shown in Fig. 7.2, which are characterized by a set of nodes or resonant frequencies, the formant frequencies. The transfer function can be approximated by a cascade combination of resonant circuits, each describing one of these nodes in the vocal tract [33]. A set of time-varying parameters is provided to control the centre frequency and bandwidth spreads of the nodes which may be activated by a voiced/unvoiced driving waveform to accurately synthesize the action of the human larynx. Formant-based synthesizers chips have been developed by NEC in their 7752c device and by Philips with the MEA8000. The possibilities of producing a set of formant frequencies for a given speech sequence as a parallel calculation has been suggested by Holmes [34] and implemented using a single chip programmable microprocessor by Quarmby and Holmes [35].

7.3 Speech recognition

Much interest has been shown during the last decade in methods used for man–machine communication, particularly communication with the computer. This is one of the most active areas for speech technology at the present time, and the most difficult. Speech recognition is concerned with pattern matching in terms of individual words separated by silences, whole phrases, or continuous speech. The process is complicated by extraneous noise, requiring methods for separating the signal from the background noise, and all three forms of speech recognition will have to deal with the problem of recognition from a variety of speakers, including those having regional accents. Early efforts were aimed at the recognition of individual words—most of these have attempted matching of the frequency spectrum of the unknown word with a set of stored spectra (templates) of a limited set of known words. Some methods stored sets of formant frequencies and zero crossing information that roughly identified the pitch of the signal. Remarkably good results, leading to a recognition accuracy of over 90 per cent, were obtained by Davis *et al.* in this way in 1952 [36]. The actual amount of hardware required for a very small vocabulary of isolated spoken words (numerical digits only) was however immense, and practical application had to await the advent of the digital computer in its compact VLSI form.

A typical isolated word recognizer using modern signal processing methods has been described by Itakura [37]. A block schematic diagram of his method is shown in Fig. 7.9. Input speech is first sent to an endpoint detector, which locates the beginning and end of an unknown utterance. After a preprocessing stage designed to limit the bandwidth and improve the SNR, an auto-correlation analysis is carried out to provide the coefficients required for LPC analysis [38]. These coding patterns are then compared with a set of similar patterns, stored as templates, for a limited vocabulary of some 50–200 words. The comparison technique is carried out dynamically as the processing proceeds, and a simularity measure accumulated against a fixed recognition threshold. Thus those template comparisons that are considerably in error are rejected at an early stage. The method admits of training for the speaker, and with several repetitions of the spoken word a reliable recognition rate of over 98 per cent can be achieved. A speaker-dependent isolated word recognition scheme has also been described by

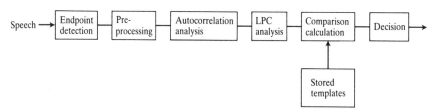

Fig. 7.9 Itakura's automatic word recognizer.

Ohga *et al.*, which uses a Walsh transform chip [39]. The speech signal is sampled at a rate of 10 kHz, and the 256 sampled data are transformed into 256 WHT coefficients. Following detection of the beginning and of the utterance (by recognition of a silence interval) the transformed segments are compared with a stored and limited set of WHT transformed words. As with the previous method a dynamic matching against a recognition threshold is used, and a *minimum distance score* [40] applied to assess the recognition result. Again a recognition rate of over 98 per cent is recorded for this scheme. Although successful by laboratory standards, these techniques still fall short of what is required in the real world of human communication. This is due to a number of practical difficulties, such as speech recording in a noisy environment (including the presence of multiple speakers), unusual accents, and above all the need to recognize continuous speech rather than isolated word utterances.

7.3.1   Digit recognition

There is a need however for a simplified but reliable scheme that can recognize the very limited vocabulary of the spoken digits 0 to 9, and much effort has been directed to this requirement. A successful technique for doing this may be based on parametric identification. A necessary preliminary is to determine the endpoints of the spoken digit. Once this has been done, the isolated digit information can be subject to parametric analysis. The parameters chosen by Rabiner and Sambur in their realization are [41]:

1. Zero crossing rate, defined as the number of zero crossings for the speech waveform in each 10 ms of the speech signal.

2. Spectral density, i.e. sum of the squared signal magnitude values taken in each 10 ms.

3. Error value and frequencies obtained from a simple LPC analysis of each 10 ms segment of the signal.

To enable a preliminary classification to be made the speech interval is segmented into three well-defined regions, a beginning transient, a middle region (vowel), and a final transient region. Each region is then characterized in terms of key features determined by the parameter values. The set of features may then be used as a template, which can be compared with a limited set of ten spoken digit templates, also in parametric form. A detailed discussion of the decision algorithm and LPC analysis is given in [1] and [42].

While isolated digit recognition can give a workable system with co-operating users (some experience in using such a system is required), a more acceptable system is one in which the digits are connected, i.e. a spoken telephone number, an identification code or in a general numeric data entry. A major problem is to determine the spoken digit endpoints. In one system, training sets of data are required from the speaker to provide statistics for the identification of voiced/unvoiced and silence speech regions. This avoids the necessity for a precise location of endpoints, thus simplifying the design problem [43].

7.3.2 Continuous speech

The goal of most speech recognition systems at the present time is to develop a linguistically oriented computer system to recognize naturally spoken phrases and sentences, i.e. continuous speech. Isolated word recognition systems enable over 99 per cent accuracy to be obtained without too much investment in terms of hardware/software. Continuous or connected speech recognition is a much more complex business and one demanding a considerable computing resource. A prime difficulty, as in isolated word recognition, is to determine where one word ends and another begins. In addition, the acoustic characteristics of connected speech displays much greater variability, depending on the context, compared with words spoken in isolation. It is no longer possible to use simple means such as template matching, due to the vast number of reference patterns that would be needed. Hence the form of analysis must be one in which the whole utterance is considered, broken down into parts, and a hierarchical system of analysis used instead of a sequential matching process.

A typical scheme of this type is the one proposed by Medress [44] for which the principle components are as shown in Fig. 7.10. The signal input is first subject to an *acoustic analysis* similar to that described above. Estimates of voice pitch, energy, and formant frequencies are made. Two further analyses are then made on a linguistic basis. A *prosodic analysis* (stress determination) component provides information

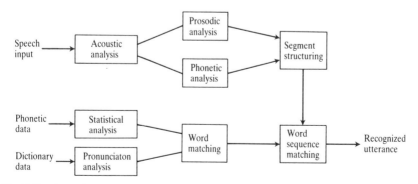

Fig. 7.10 A continuous speech recognition system.

about the syllabic structure—where the syllables are placed, and estimates of their stress. A *phonetic analysis* indicates the location and type of stops, sibilants, vowels, fricatives, etc. in the speech signal. Finally, these various pieces of analysis information are separated into segments for matching operations with a phonetic database and a set of dictionary pronunciations are used to identify the spoken words. A realization of this complex scheme has been carried out by Sperry Univac [45, 46]. Limiting the number of male talkers to three, the technique achieves an average phrase recognition accuracy of about 94 per cent and an average word recognition accuracy of about 98 per cent.

The amount of processing required is however quite high, and speed of operation is considerably slower than real-time. Parallel processing techniques have been suggested to enable real-time operation to be realized [46].

7.3.3   Word spotting

An adjunct to continuous speech analysis is an automatic technique for isolated word recognition [47, 48]. The object is to identify individual words out of an input speech stream independently of the spoken dialect. One use for the technique is to train the user to speak very distinctly to minimize co-articulation problems, but there are obviously very many other uses.

The basic elements are similar to those shown in Fig. 7.10. Acoustic, prosodic, and phonetic analysis of the speech signal are carried out, together with a segmental structuring of the data to provide a linear sequence of separate segments suitable for matching from keywords contained in a database. Actually several sorts of matching are carried out sequentially, and this increases the complexity over the previous method. Matching is carried out in terms of individual word phonetics, dictionary pronunciation, and spectral coefficients, with the spectra of the words using contained in the database. Each form of comparison carries with it its own 'scoring' for the matches obtained and the final selection made on a statistical basis.

7.4   Data communication

Transmission of speech (telephone conversations) and more recently digital data is now well established using a network of line communication employing *frequency-division-multiplexing* (FDM) and *single-sideband signals* (SSB) as the principle means of accommodating the very large numbers of transmission channels [49, 50].

Let us consider first the modulating process. An input analog signal, $x(t) = \cos \omega_m t$ is modulated by a carrier signal, considered here as a pure cosinusoidal signal, $z(t) = \cos \omega_c t$ to give a composite signal,

$$y(t) = \cos \omega_m t \cos \omega_c t$$
$$= 1/2[\cos(\omega_c + \omega_m)t + \cos(\omega_c - \omega_m)]. \qquad (7.15)$$

The two angular frequencies generated

$$(\omega_c + \omega_m) \quad \text{and} \quad (\omega_c - \omega_m) \qquad (7.16)$$

are the upper and lower *sidebands* of the double-sideband modulated (DSB) signal $y(t)$. One way to generate a SSB signal would be to pass the DSB signal through a bandpass filter to remove the undesired sideband. However, it would be extremely difficult to design a filter having the correct frequency characteristics and a more realistic solution is to use a phase shift method.

Assuming that we wish to transmit the lower sideband, $\cos(\omega_c - \omega_m)$, then this can be written

$$\cos(\omega_c - \omega_m) = 1/2[\cos \omega_c t \cos \omega_m t + \sin \omega_c t \sin \omega_m t]. \qquad (7.17)$$

By introducing a $\pi/2$ phase shift into the cosine waveforms, a cancelling of the upper sideband can be achieved by the arrangement shown in Fig. 7.11. A multiplexed SSB is obtained by summation of the SSB outputs of each of the modulated channel signals. This is translated to the required carrier frequency by a further modulating process (which can also be a SSB one). The complete system for a multiplexer using SSB modulation is shown in Fig. 7.12.

Recent trends in digital exchanges and communication networks has resulted in the growth of additional communication systems employing *time-division-multiplexing* (TDM), which handles digital data directly without the need to modulate a high-frequency carrier signal. Figure 7.13 illustrates in simplified form the principles of TDM. In order to transmit continuous analog signals over such a system, it is necessary first to sample the signal and then to encode each sample into a unique binary code. The principle way to carry this out is to use *pulse-coded modulation*

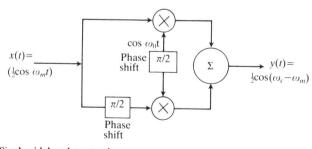

Fig. 7.11 Single-sideband generation.

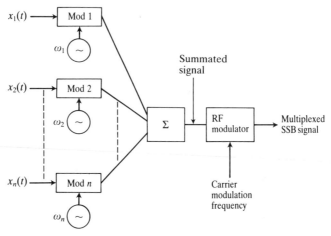

Fig. 7.12 Frequency-division multiplexing.

(PCM), discussed in relation to speech encoding during the earlier part of this chapter.

TDM systems are gradually supplanting the earlier binary-coded analog FDM carrier systems over the telephone networks for reasons of efficiency and economy (quite apart from the need to accommodate an increasing amount of digital data traffic). This process will however take some time to complete, and in the meantime it is necessary to be able to convert from one system to another. The device used to carry this out is the *transmultiplexer,* which will be considered later.

Transmission of digital data over the older FDM systems is accomplished through a data *modem* designed to modulate serial digital data onto an FDM carrier before transmission and to demodulate the signal at the receiving end to recover the digital datastream. There has been some interest in parallel transmission methods in which each member of a digit sequence modulates a single-tone subcarrier and the digital signal is thus transmitted as an FDM parallel data signal [51]. The advantage of doing

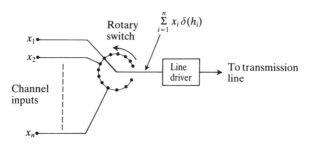

Fig. 7.13 Time-division multiplexing.

this is that equalization is avoided, the effect of impulsive noise is reduced, and better use is made of the available bandwidth. For large numbers of channels the arrays of sinusoidal generators and coherent demodulators required can make the system expensive to realize. It can be shown however that the multitone data signal is effectively the Fourier transform of the original serial data sequence so that it is necessary only to transmit the transformed signal, which may then be demodulated at the receiver and through an inverse transformation [52]. This enables a simpler system to be evolved, making use of fast digital transformation and digital signal processing techniques. Examples may be found in [52, 53] for modem use and in [54] for single-sideband digital FDM application.

7.4.1 Data compression

Nearly all signals contain a certain amount of redundant information, which may be removed to permit the basic information only to be transmitted in compressed form. A number of redundancy removal schemes have been proposed and used, the most well-known being:
  (a) source symbol encoding;
  (b) run-length encoding;
  (c) predictive encoding; and
  (d) transform methods.
A detailed mathematical description of these and other methods may be found in [55–59]. In most data compression schemes their effectiveness is described in terms of a *data compression ratio* (DCR)

$$d = \frac{\text{bits in input sequence}}{\text{bits in output sequence}} \qquad (7.18)$$

for the same message or signal. The effectiveness of several data compression methods depends on the statistical independence of the pieces of data in the signal. Clearly, for a signal highly dependent on previous signals in a series, such as encoded speech or frame-by-frame television transmission, higher data compression ratios can be achieved. In source symbol encoding attention is directed towards developing binary codes to represent the numbers being transmitted having the smallest sequence of binary bits [56, 60].

Run-length encoding and predictive encoding are both methods that rely on the statistics of the data being transmitted. With run-length encoding reliance is placed on the fact that redundant sources often produce nearly constant signal levels, constituting a 'run' of the signal (e.g. a series of frames for a television transmission). Compression is achieved by only transmitting the initial signal levels followed by information on the length of the run and any level changes that may have occurred. Predictive coding is a rather more sophisticated approach in which a set of previous data samples is used to predict the next set of data samples. Compression is then achieved by excluding the samples that can be predicted in this way from the data being transmitted. We considered a specialized application of this technique in Section 7.2.7 in connection with speech encoding.

Transform coding techniques have probably been the most successful and widely applied of all these methods of data compression (see Section 8.2.1). In these techniques the data is converted into the transform domain, the number of transformed coefficients reduced, and the compressed data transmitted or stored. Reconstruction of the compressed data is obtained by first replacing the coefficients not transmitted by zeros and then carrying out an inverse transformation and low-pass filtering. All the orthogonal transforms discussed in Chapters 4 and 5 have been used in this form of data compression for both single and multi-dimensional data. Table 7.2 shows the compression ratios achieved for image data using a number of different orthogonal transformations (with acknowledgement to Dr. J. O. Thomas [61]). Selection of transform coefficients for reduction is frequently made using a threshold filtering procedure in which coefficients occupying the highest frequency or sequency position are reduced to zero. Other more efficient methods

Table 7.2 Data compression ratio using various unitary transforms (after J. O. Thomas [61])

Transform	DCR	MSE
FWT	4:1	0.31
	6:1	0.55
	12:1	1.25
FWT	4:1	0.22
	6:1	0.43
	12:1	1.02
FHT	4:1	0.19
	6:1	0.35
	12:1	0.85
FST	4:1	0.16
	6:1	0.30
	12:1	0.67
KLT	4:1	0.14
	6:1	0.26
	12:1	0.57

have been used, including discriminant analysis [62], principle component analysis [63], maximum likelihood, and various parametric and non-parametric methods [61]. These will not be considered further here.

7.4.2 Sequency multiplexing

An outline of the principle of FDM systems was given at the beginning of this section. Less well known are the multiplexing methods that have been applied using the Walsh transform as the transmission carrier in place of a sinusoidal carrier [64]. These are known as *sequency-division-multiplexing* (SDM) systems. One of the advantages of using the Walsh functions as carriers, apart from their ease of generation through digital logic, is that the multiplication process produces only one side-band and not two, as is the case with sinusoidal products. The reason is that the Walsh functions form a group under multiplication so that the product of two Walsh functions having sequency values n and m is a further Walsh function having a single sequency equivalent to the dyadic sum, $n \oplus m$, i.e.

$$\text{WAL}(n, t)\text{WAL}(m, t) = \text{WAL}(n \oplus m, t). \tag{7.19}$$

This may be compared with eqn (7.15) for FDM, which shows two different frequency signals, produced simultaneously through the product of $\cos \omega_n t$ and $\cos \omega_m t$, so that filtering or some other process is needed to produce a single-sideband signal. The use of a Walsh function as a carrier thus permits sideband filters to be omitted and also simplifies implementation using integrated circuit technology. A number of experimental analog SDM systems have been built and are described elsewhere. A notable early design is a working system designed for 256 voice channels by Hubner [65] of the West German Post Office. Other systems are discussed by Bagdasarjanz and Loretan [66] and a review paper given by Schreiber [67].

Use of Walsh functions for continuous multiplexing systems have been limited by problems in achieving accurate synchronization, finite rise time of the Walsh carriers, and insufficiently linear amplifiers in the communication system. Rather more promising results have been achieved through *digital sequency division multiplexing systems* (DSDM) designed to carry digital traffic.

A number of advantages of DSDM over FDM are noted by Harmuth and Murty [68]. The logic circuits used are small and highly reliable and no setting-up operations are required. There is also the immunity to burst-type interference arising from the correlation process of demodulation, which is not possible with TDM. Several methods of DSDM have been proposed. One of the most successful of these is the *majority logic multiplexing* system described by Gordon and Barrett [69]. This is a digital multiplexing system in which only the sign of the multiplexed signal is transmitted. This is possible, since with binary signalling it is only necessary to determine the sign of the correlation coefficient for the transmitted information, relative to a single channel, for this to be unambiguously recovered. Each channel of binary data to be multiplexed is amplitude-modulated, with one of a set of binary Walsh codes acting as the transmission carriers. The summated signal is passed through a hard limiting device that transmits only the sign of the majority logical value in each time slot. Demultiplexing is carried out by finding the correlation coefficient of the transmitted signal with each of a set of replica Walsh codes generated at the receiver. The number of channels that may be used with this system depends on the existence of a matrix, the rows of which show a sign-invariant correlation coefficient after being modulated, summed, and threshold-limited. It can be shown that such a matrix can be formed from the Walsh functions $WAL(1, t)$ to $WAL(7, t)$ [69]. The use of larger matrices to increase the number of channels has been described by Mukherjee [70].

A number of majority logic multiplexors of this type have been constructed for digital telephone signals [71]. A particularly economic system in terms of bandwidth required is the adaptive Walsh multiplex scheme described by Zheng and Youwei [72] and implemented as a 12-channel digital telephone system.

7.4.3 TDM–FDM conversion

With the increasing use of PCM in the newer digital telephone exchanges and the slow rate of replacement for existing analog FDM equipment, there is a need for translating hardware to permit communication between the two systems. This has resulted in a number of designs for *transmultiplexors,* i.e. TDM to FDM and FDM to TDM converters operating on a multichannel basis. Generally the TDM scheme used is PCM, although some waveform coding in adaptive PCM (APCM) and adaptive delta modulation (ADM) is also found [73].

To perform the conversion from PCM to FDM each channel has first to be single-sideband modulated with its appropriate carrier frequency using a sampling rate interpolated up to the higher rate demanded by the standard telephone group frequencies [74]. An early method for achieving this using digital hardware has been described by Freeny *et al.* [75]. This uses a digital version of a classical *Weaver modulation,* shown in simplified form for a single channel in Fig. 7.14 [76]. A PCM input signal sampled at 8 kHz is modulated by a 2 kHz carrier along two paths, one involving a sine waveform modulator and known as the in-phase path, and the other by a cosine waveform modulator and called the quadrature path. The modulated signal on each path is subject to low-pass digital

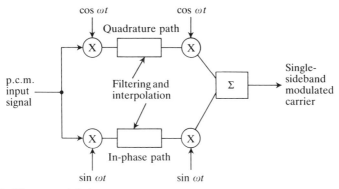

cos ωt cos ωt

Fig. 7.14 Weaver modulation.

filtering and interpolated up to 112 kHz sampling rate. At this point the signal on each path is again modulated by an in-phase and quadrature carrier. Addition of the modulated signals results in the cancelling of the lower sideband, producing the required SSB signal.

Using an array of such circuits sharing a common summing circuit a FDM signal will become available at the output. It has been shown by Darlington [77] that this process of modulating and demodulating a group of channels can be carried out by algorithms somewhat similar to the FFT providing that the number of channels is a power of two. Further work by Bellanger and Daguet [78] showed that the translation problem may be tackled directly by using an FFT in conjunction with a polyphase network. This is understandable if we consider the FDM signal as a frequency domain representation of some kind of conglomerate for a serial time signal; in this case a PCM representation. Similarly we can regard the demultiplexing of the FDM signal as a spectral analysis operation. In an implementation due to Bonnerot *et al.* [79] a special form of the FFT is used, called the *odd-time odd-frequency discrete Fourier transform* (O^{2}DFT) [80]. This is an efficient algorithm obtained by taking the time samples of a symmetric real-valued series as odd multiples of half the sampling period h and frequency samples as odd multiples of $1/2Nh$. The definition of a Fourier transform pair given for the O^{2}DFT is

$$X_n = \frac{1}{N} \sum_{i=0}^{N-1} x_i \exp[-2\pi j(2n+1)(2i+1)/4N] \qquad (7.20)$$

$$x_i = \sum_{n=0}^{N-1} X_n \exp[2\pi j(2n+1)(2i+1)/4N] \qquad (7.21)$$

where $i, n = 0, 1, \ldots, N-1$.

It is shown in [79] that in this form the transform can be used to realize fairly easily a bank of bandpass filters having adjacent frequency bandwidths and a high cut-off rate. The number of multiplications required is less than earlier methods and the need for elaborate complex signals required with the Weaver modulation technique may be avoided. Since the form of the O^2DFT and its inverse is identical, the same transform computation can operate in both a multiplexing and demultiplexing operation.

A further motivation behind decomposition of the PCM signal through the use of a DFT is of course to make use of an efficient FFT algorithm. The efficiency of this translation may be further improved by using the reduced multiplication algorithms considered in Chapter 4. One such scheme is described by Peled and Winograd, which reduces the number of multiplications by a factor of ten in exchange for greater program complexity [81]. Similar improvements have also been obtained using the FCT [82]. (The reader may have noticed that this technique of using the FFT as a method of simulating a filter bank has also been used to implement the sub-band coder for speech compression [82]. A general description of the method is given in [83]).

References

1. Rabiner, L. R. and Schafer, R. W. *Digital processing of speech signals.* Prentice-Hall, Englewood Cliffs (1978).
2. Rabiner, L. R. and Gold, B. *Theory and applications of digital signal processing.* Prentice-Hall, Englewood Cliffs (1975).
3. Peled, A. and Liu, B. *Digital signal processing, theory, design and implementation.* Wiley, New York (1976).
4. Babu, B. N. S. Performance of an FFT-based voice coding system. *IEEE Trans. Acoust. Speech Signal Process.* **ASSP-31** (5), 1320–8 (1983).
5. Berlekamp, E. R., ed. *Key papers in the development of coding theory.* IEEE Press, New York (1974).
6. Richards, M. A. Helium speech enhancement using the short-time Fourier transform. *IEEE Trans. Acoust. Speech Signal Process.* **ASSP-30** (6), 841–53 (1982).
7. Flanagan, J. L., Coker, C. H., Rabiner, L. R., Schafer, R. W., and Umedu, N. Synthetic voices for computers. *IEEE Spectrum* **7,** 22–45 (1970).
8. Rabiner, L. R. and Schafer, R. W. Digital techniques for computer voice response; implementation and application. *IEEE Proc.* **64** (4), 416–33 (1976).
9. Flanagan, J. L. *Speech analysis, synthesis and perception* (2nd edn). Springer, New York (1972).
10. Gold, B. and Rabiner, L. R. Analysis of digital and analog formant synthesis. *IEEE Trans. Acoust.* **AU-16,** 81–94 (1968).
11. Barnwell, T. P., Bush, A. M., Neal, J. B., and Stroh, R. W. Adaptive differential PCM speech transmission. RADC-TR-74-77, Rome Air Development Centre, July (1974).

12. Schindler, H. R. Delta modulation. *IEEE Spectrum* **7**, 69–78 (1970).
13. Pornoff, M. R. Implementation of the digital phase vocoder using the FFT. *IEEE Trans. Acoust. Speech Signal Process.* **ASSP-24**, 243–8 (1976).
14. Flanagan, J. L., Schroeder, M. R., Stal, B. S., Crochiere, R. E., Jayant, N. S., and Tribolet, J. M. Speech coding. *IEEE Trans. Commun.* **COM-27** (4), 710–37 (1979).
15. Zelinski, R. and Noll, P. Adaptive transform coding of speech signals. *IEEE Trans. Acoust. Speech Signal Process.* **ASSP-25** (4), 299–309 (1977).
16. Cheung, J. Y. and Holden, A. D. C. Computer modelling and estimation of linguistic stress patterns. IEEE International Conference on Acoustics, Speech and Signal Processing, Piscataway, NJ (1976).
17. Holden, A. D. C. Speech processing. In *Digital waveform processing and recognition* (ed. C. H. Chen) pp. 92–102. RCR Press, Florida (1982).
18. Dudley, H. The vocoder. *Bell Lab. Rec.* **17**, 122–6 (1939).
19. Schroeder, M. R. Vocoders; analysis and synthesis of speech. *IEEE Proc.* **54**, 720–34 (1966).
20. Tufts, D. W., Levinson, S. E., and Rao, R. Measuring pitch and formant frequencies for a speech understanding system. IEEE International Conference Acoustics, Speech and Signal Processing, pp. 314–317, Piscataway, NJ (1976).
21. Rabiner, L. R., Cheng, G. M. J., Rosenberg, A. E., and McGonegan, C. A. A comparative performance study of several pitch detection algorithms, *IEEE Trans. Acoust. Speech Signal Process.* **ASSP-24** (5), 399–418 (1976).
22. Sondhi, M. M. New methods of pitch extraction. *IEEE Trans. Acoust. Electroacoust.* **AU-16** (2), 262–6 (1968).
23. Noll, A. M. Cepstrum pitch determination. *J. acoust. Soc. Am.* **41**, 293–309 (1967).
24. Elias, P. Predictive coding. *IRE Trans. Inf. Theory* **IT-1**, 16–33 (1955).
25. Markel, J. D. The SIFT algorithm for fundamental frequency estimation. *IEEE Trans. Audio Electroacoust.* **AU-20** (5), 367–72 (1972).
26. Makhoul, J. Linear prediction: a tutorial review. *IEEE Proc.* **63**, 561–80 (1975).
27. Markel, J. D. and Gray, A. H. *Linear prediction of speech*. Springer, Berlin (1976).
28. Oppenheim, A. V. A speech analysis/synthesis system based on homomorphic filtering. *J. acoust. Soc. Am.* **54**, 458–65 (1969).
29. Bogert, B. P., Healey, M. J., and Tukey, J. W. The quefrequency analysis of time series for echoes; cepstrum, pseudo-autocovariance, cross spectrum and saphe cracking. In *Applied time series* (ed. D. Findley). Academic Press, New York (1963).
30. Hofstetter, E. M., Tierney, J., and Wheeler, O. Microprocessor realization of a linear predictor vocoder. *IEEE Trans. Acoust. Speech Signal Process.* **ASSP-25** (5), 379–87 (1977).
31. Quarmby, D. J., ed. *Signal processor chips*. Granada, London (1984).
32. Holmes, J. N., Mattingly, I. G., and Shearme, J. N. Speech synthesis by rule. *Language Speech* **7**, 127–43 (1964).
33. Steves, K. N. and Faut, G. C. M. An electrical analog of the vocal tract. *J. acoust. Soc. Am.* **25**, 734–42 (1953).
34. Holmes, J. N. Parallel formant vocoders. IEEE Eascon Conference Proceedings, Washington, pp. 713–18 (1978).

35. Quarmby, D. J. and Holmes, J. N. Implementation of a parallel formant speech synthesizer using a single-chip programmable signal processor. *IEE Proc.* **131F** (6), 563–9 (1984).
36. Davis, K. H., Biddulph, R., and Balashek, S. Automatic recognition of spoken digits. *J. acoust. Soc. Am.* **24** (6), 637–42 (1952).
37. Itakura, F. Minimum prediction residual principle applied to speech recognition. *IEEE Trans. Acoust. Speech Signal Process.* **ASSP-23** (1), 67–72 (1975).
38. Markel, J. D. and Gray, A. H. A linear predictor vocoder simulation based upon the autocorrelation method. *IEEE Trans. Acoust. Speech Signal Process.* **ASSP-22** (2), 124–34 (1974).
39. Ohga, H., Yabuche, H., Tsuboka, E., Mayami, K., Adachi, K., and Nishijiwa, O. A Walsh–Hadamard transform LSI for speech recognition. *IEEE Trans. consumer Electron.* **CE-28** (3), 263–69 (1982).
40. Sakoe, H. and Chiba, S. A dynamic programming approach to continuous speech recognition. Proceedings of an International Congress on Acoustics, Budapest, Hungary pp. 200–13 (1971).
41. Rabiner, L. R. and Sambur, M. R. Some preliminary results on the recognition of connected digits. *Bell Syst. tech. J.* **54** (2), 297–315 (1975).
42. Makhoul, J. and Wolf, J. The use of a two-pole linear prediction model in speech recognition. Report No. 2537, Bold Berenek and Newman, Cambridge, MA (1973).
43. Rabiner, C. R. and Sambur, M. R. Some preliminary results on the recognition of connected digits. *IEEE Trans. Acoust. Speech Signal Process.* **ASSP-24** (2), 170–82 (1976).
44. Medress, M. F., Skinner, T. E., Kloker, D. R., Dillier, T. C., and Lea, W. R. A system for the recognition of spoken connected word sequences. IEEE Conference on Acoustics, Speech and Signal Processing. IEEE Catalogue No. 77 CH1197-3 ASSP, pp. 468–73 (1977).
45. Skinner, T. E. Toward automatic determination of the sounds comprising spoken words and sentences, Sperry-Univac Report No. PX12124 (1977).
46. Skinner, T. E. Speaker invariant characterization of vowels, liquids, and glides using relative formant frequencies. *J. acoust. Soc. Am.* **62** (S1(A)) 55 (1977).
47. Medress, M. F., Dillier, T. C., Kloker, D. R., Lutton, L. C., Oredson, H. N., and Skinner, T. E. An automatic word spotting system for conversational speech. IEEE International Conference on Acoustics, Speech and Signal Processing. IEEE Catalogue No. 78 CH1285-6, pp. 712–17 (1978).
48. White, G. M. and Sambur, M. R. Speech recognition research at ITT defense communication division. In *Trends in speech recognition* (ed. W. A. Lea). pp. 469–82. Prentice-Hall, Englewood Cliffs (1980).
49. Lucky, R. W., Salz, J., and Weldon, J. R. *Principles of data communication.* McGraw-Hill, New York (1968).
50. Lathi, B. P. *Modern digital and analog communication systems.* Holt, Rinehart, and Winston, New York (1983).
51. Chang, R. W. and Gibby, R. A. A theoretical study of the performance of an orthogonal multiplexing data transmission scheme. *IEEE Trans. commun. Technol.* **COM-16,** 529–40 (1968).
52. Weinstein, S. B. and Ebert, P. M. Data transmission by frequency division

multiplexing using the discrete Fourier transform. *IEEE Trans. commun. Technol.* **COM-19** (5) 628–34 (1971).

53. Powers, E. N. and Zimmerman, M. S. TADIM—a digital implementation of a multichannel data modem. *IEEE int. Conf. Commun.* Philadelphia (1968).

54. Maruto, R. and Tomozama, A. An improved method for digital SSB–FDM modulation and demodulation. *IEEE Trans. Commun.* **COM-26,** 720–5 (1978).

55. Winter, P. A. Transform picture coding. *IEEE Proc.* **60,** 809–20 (1972).

56. Huffman, D. A. A method for the construction of minimum redundancy codes. *IRE Proc.* **40,** 1098–101 (1952).

57. Babibi, A. Survey of adaptive coding techniques. *IEEE Trans. Commun.* **COM-25,** 1275–84 (1977).

58. Turner, L. F. Data compression techniques as a means of reducing the storage requirements for satellite data. *Radio electron. Engng.* **43,** 599–608 (1973).

59. Berger, T. *A mathematical basis for data compression in rate distortion theory.* Prentice-Hall, Englewood Cliffs (1971).

60. Shannon, C. F. A mathematical theory of communication. *BTSJ* **27,** 623–56 (1948).

61. Thomas, J. O. Digital imagery processing. In *Issues in digital image processing* (ed. R. M. Haralick and J. C. Simon) pp. 247–90. Noordhoff, Amsterdam (1980).

62. Roper, R. B. Analytical and interactive technique for multivariate data compression and classification. AFCRL Report No. 0540 (1979).

63. Hotelling, H. Analysis of a complex of statistical variables into principle components. *J. educ. Psychol.* **24,** 417 and 498 (1933).

64. Beauchamp, K. G. *Applications of Walsh and related functions.* Academic Press, New York (1984).

65. Hubner, H. Analog and digital multiplexing by means of Walsh functions. Proceedings of a Symposium on the Applications of Walsh Functions, Washington, DC AD727000, pp. 180–191 (1971).

66. Bagdasarjanz, F. and Loretan, R. Theoretical and experimental studies of a sequence multiplex system, Proceedings of a Symposium on the Applications of Walsh Functions, Washington, DC, AD707431, pp. 36–40 (1970).

67. Schreiber, H. H. A review of sequency multiplexing, Proceedings of a Symposium on the Applications of Walsh Functions, Washington, DC, AD763000, 18–33 (1973).

68. Harmuth, H. F. and Murty, S. S. R. Sequency multiplexing of digital signals. Proceedings of a Symposium on the Applications of Walsh Functions, Washington, DC, AD763000, pp. 202–9 (1973).

69. Gordon, J. A. and Barrett, R. Correlation recovered adaptive majority multiplexing. *IEEE Proc.* **118,** 417–22 (1971).

70. Mukherjee, A. K. and Mukhopadhyay, D. A method for increasing the number of majority multiplexed channels. *IEEE Proc.* **66,** 1096–7 (1978).

71. Qishan, Z. Walsh telemetry systems. *Telemetry Engng.* **1** (3) 8–11 (1980).

72. Zheng, H. and Youwei, Y. An adaptive multiplexed delta modulation system. IEEE Symposium Electromagnetic Compatibility, Baltimore, pp. 346–8 (1980).

73. Jeong, H. and Un, K. C. A PCM/ADM and ADM/PCM code converter. *IEEE Trans. Acoust. Speech Signal Process.* **ASSP-27** (6), 762–8 (1979).
74. *Handbook of Data Communication,* NCC, Manchester, England (1975).
75. Freeny, S. L., Kieburtz, R. B., Mina, K. V., and Tewksbury, S. K. System analysis of TDM–FDM translator/digital *A*-type channel bank. *IEEE Trans. commun. Technol.* **COM-19** (6), 1050–9 (1971).
76. Weaver, D. K. A third method of generation and detection of single sideband signals. *Proc. IRE* **44,** 1703–5 (1956).
77. Darlington, S. On single sideband modulators, *IEEE Trans. circuit Theory* **CT-17** (3), 400–414 (1970).
78. Bellanger, M. and Daguet, J. L. TDM–FDM translator; digital polyphase and FFT, *IEEE Trans. Commun.* **COM-22,** 1199–204 (1974).
79. Bonneret, G., Condreuse, M., and Bellanger, M. G. Digital processing techniques in the 60ch, transmultiplexer. *IEEE Trans. Commun.* **COM-26** (5), 698–707 (1978).
80. Bonnerot, G. and Bellanger, M. G. Odd-time odd-frequency discrete Fourier transform for symmetric real-valued series. *IEEE Proc.* **64** (3), 392–3 (1976).
81. Peled, A. and Winograd, S. TDM–FDM conversion requiring reduced computation complexity. *IEEE Trans. Commun.* **COM-26** (5), 707–19 (1978).
82. Narasimha, M. J. and Peterson, A. M. Design of a 24ch transmultiplexer. *IEEE Trans. Acoust. Speech Signal Process.* **ASSP-27** (6), 752–61 (1979).
83. Narasimha, M. J. and Peterson, A. M. Design and application of uniform digital bandpass filter banks. IEEE International Conference on Acoustics, Speech and Signal Processing, Tulsa (1978).

8

SONAR, SEISMOLOGY, AND RADAR

8.1 Introduction

Three application areas having much in common from the signal
processing point of view are sonar, seismology, and radar. All are
concerned with the detection of signals propagated or transmitted
through a medium for the purpose of signal source or reflected signal
source identification and description. In the sonar case the medium is
water, restricting the frequencies used to the upper end of the audio
range and constraining the signal velocities to about 1.5 km s^{-1}. Seismic
signals on the other hand propagate through the earth's crust and occupy
a frequency range of only a few Hz, with a variable rate of velocity
considerably less than that of sound in air. At the other end of the
frequency range radar signals propagate at light velocity and utilize a
band of frequencies extending into hundreds of MHz.

Despite these differences the signal processing methods used share
many common characteristics and tackle similar problems (e.g. scatter
and noise degradation). One consequence of this is that the three
disciplines borrow freely of each other's methods and problem solutions
and in this sense can be regarded as interdisciplinary studies.

In the necessarily limited space of this chapter only a superficial view of
these three applications can be taken, with attention focused on
techniques of analysis employing transformation, particularly where these
are shared by all three topics. For the reader requiring either a more
extensive background in one or more of these topics or further details of
analysis methods, the literature is extensive and a number of key
references are included at the end of this chapter. In addition, current
research is published in several journals including the relevant transac-
tions of the IEEE and specialist journals such as the *Journal of the
Acoustical Society of America* (for sonar) and *Geophysics* (for
seismology).

8.2 Sonar

The essential feature of sonar (SOund NAvigation and Ranging) systems
is the propagation of a signal towards the object to be detected, (the
target), and deductions made concerning its exact direction, distance, and
sometimes its spatial characteristics from reflected signals travelling along

the reverse path of propagation. The signal can comprise a single pulse, a train of pulses or a frequency-modulated waveform.

The level of the signal returned can be quite high. The sea is a better medium for sound transmission than air and supports a transmission velocity some 4.5 times greater. This facilitates the measurement of the echo characteristics, sometimes at considerable distances from the source of the signal. This velocity also happens to be about 100 times faster than the maximum speed of ships, which is ideal for Doppler frequency shift measurement [1]. The frequencies employed are convenient for electronic design and processing and for the most part fall between 10 Hz and 20 kHz. Higher frequencies up to 500 kHz are also used for short-range high definition systems. Above 50 kHz however absorption during transmission from various causes increases at a rate proportional to f^2, thus reducing system sensitivity considerably.

The difficulties in sonar processing lie in the restrictive nature of the communication channels between the sound source and receiver. The inhomogeneities in the volume of the water and roughness of the surface and sea bottom boundaries introduce various types of signal distortion and multipath effects. In addition the detector and measured target are likely to be in relative motion and since the target may be large it can introduce a scattering effect in the returned signal. This latter will also be the case for extraneous moving bodies located between the detector and the target (e.g. shoals of fish), which contribute a measure of noise to the returned signal. All of these unwanted effects complicate the operations involved in optimal signal processing, so that fairly complex processing becomes necessary [2–5].

Sonar technology may be divided into two fields:

1. *Passive sonar,* which is concerned with the detection of signals radiated directly from the source, e.g. the noise of a ships' propellor.

2. *Active sonar,* which relates to the detection of a secondary echo signal reflected from an underwater target. Here the location and duration of the primary signal source is under the control of the operator. This is a major method used in sonar for the location of moving underwater targets.

8.2.1 Passive sonar

The object of underwater passive listening is to detect the presence of noise sources. Since the structure of these noise signals is unknown the main criterion that can be used for analysis is their spatial coherence. Two or more measurements are therefore taken at different detector locations. The signals actually reaching the detector(s) are almost identical except for a time delay according to their direction and distance from the source. The signals will be blurred by other coherent noise sources such as echoes from other underwater objects, noise-emitting sources and a general background noise relating to the in-homogeneous transmission paths. There is a major difference in opera-tion between detection in shallow and deep water since the former will exhibit more variability in propagation and reverberation and probable multiple reflections from the spatially inhomogeneous bottom and surface boundaries [6].

Reflections from the surface and sea bottom are however valuable for the measurement of target range and direction. A simplified model of the propagation between an underwater noise source S and a receiver X is shown in Fig. 8.1. Neglecting for the moment relative movement between S and X, the position of X in relation to the sea surface, h_1 and sea bottom, h_2 will be known. If we are able to measure the delay, D_1 between the direct and surface reflected signal paths, R and $R_{1s} + R_{2s}$ or D_2 between the direct and bottom reflected paths, R and $R_{1b} + R_{2b}$ then it is possible to represent these delays in terms of R as

$$D_1 - ((R^2 + 4h_1^2 - 4h_1 S_a)^{1/2} - R)/V \tag{8.1}$$

and

$$D_2 = ((R^2 + 4h_2^2 - 4h_2 S_a)^{1/2} - R)V \tag{8.2}$$

where S_a is the altitude of the source S above the plane of the receiver X, and V is the velocity of the signal.

Given two successive measurements for eqns (8.1) or (8.2), V can be determined.

An alternative method of using a time delay is to note the time delay between the same noise source received at two sensors separated by a known distance, d (Fig. 8.2). In either case the problem is to measure this time delay automatically from the two measured signals, which we will call x_1 and x_2.

A common method is to use the cross-correlation function

$$R_{x1,x2}(\tau) = 1/(T - \tau) \int_{t=\tau}^{T} x_1(t)x_2(t + \tau)\, dt \tag{8.3}$$

where T represents the observation interval.

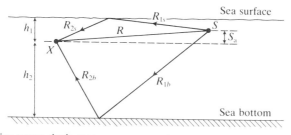

Fig. 8.1 Passive sonar: single sensor measurement.

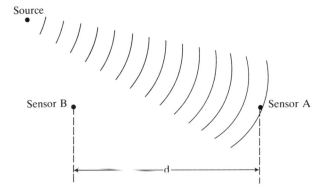

Fig. 8.2 Passive sonar: double sensor measurement.

The maximum value of $R_{x1,x2}$ for a range of delays (τ) will give the required estimate of delay D. A series of such measurements allows R, S_a and V to be obtained for the noise source moving in relation to the receiver at X. In a practical evaluation the cross-spectral density is first determined through the FFT, and the cross-correlation function is obtained through the Wiener–Khintchine relationship [7] (see Chapter 1). The chirp-Z transform has also been used [8]. The presence of noise will modify this process and various methods of minimizing this, including particularly the use of a Kalman filter, are described elsewhere [9, 10].

Cepstrum techniques have also been studied extensively for underwater signal detection and time delay estimation in the presence of noise [11–13]. Since only the time of the echo is required the power cepstrum given in eqn (1.36) can be applied, which does not retain the phase information about the signal. The composite signal $y(t)$ is assumed to represent an overlapping of the two signals, $x(t)$ and a delayed and attenuated replica, $ax(t - \tau)$, together with a stationary noise factor $n(t)$, i.e.

$$y(t) = x(t) + ax(t - \tau) + n(t). \qquad (8.4)$$

The Fourier transform $Y(\omega)$ of $y(t)$ is taken to yield,

$$Y(\omega) = X(\omega)[1 + a\exp(-\mathrm{j}\omega\tau)] + N(\omega). \qquad (8.5)$$

The logarithm of the squared value of $Y(\omega)$ is then obtained,

$$K_y(\tau) = \log_\mathrm{e} |Y(\omega)|^2. \qquad (8.6)$$

This is the power cepstrum for the composite signal and acts to separate out the contribution of $x(t)$, which will be concentrated near to $t = 0$, and the echo which gives a dominant peak corresponding to the time delay, τ. Due to the presence of a noise spectrum, $N(\omega)$ some smoothing is generally necessary to permit a clear identification of the peak value, which then forms the estimated value of the time delay [12].

8.2.2 Sonar arrays

Sonar target range and bearing using a linear array of hydraphone detectors (usually situated along the sea bottom), is a well-used technique which does not rely on reflections from the boundary surfaces. The sonar target is assumed to generate a wide-band zero mean noise source and the sensor outputs are measured for a time interval which is long compared with the time required for the signal wavefront to propagate across the array. As discussed later in connection with seismic arrays (Section 8.3.2) the maximum directional sensitivity of the array is dependent on introducing a set of phased delays into the outputs of the array elements before the array signals are added together. Calculation of an optimum set of delays for the noise signal being received is known as *beam forming* or steering for the array.

As in two-element detection described above the object is to determine the delays between signals received at the various elements along the linear array so that the bearing and distance of the target can be estimated. A multiple cross-correlation method is frequently applied using eqn (8.3). Given M hydraphone elements a set of $M(M-1)/2$ cross correlations is used to form all of the cross correlograms corresponding to all of the M detector outputs taken two at a time. The positions of the peaks $R_{x1,x2}$ of the cross-correlogram are used as estimates for the inter-element delays. These steering delays may be introduced into the individual signals from the array in order to enhance the summed and averaged signal from the complete array. Methods of analysis for determining range and the direction of the hydrophone array signals are given in [14].

8.2.3 Active sonar

With active sonar the user generates and directs the transmission of sound energy so that the signals employed may be arranged to match the acoustic environment for his requirement. Early systems simply consisted of a piezo-electric transducer excited near its resonant frequency, which acted as the transmitter for the sound waves. The portion of the energy returned from the target was detected using a narrow band energy detector. Later developments include programmable adaptive modulators, multi-element transducer arrays capable of directing the sound beam in a given direction, and a number of complex digital signal processing techniques designed to improve the SNR of the signal and to assess automatically the direction, location, and structure of the target.

Three commonly used signals are illustrated in Fig. 8.3. The first comprises a burst of sinusoidal energy having a duration of a few ms up to a second or so, with centre frequencies contained within the audio range. The second and third are modulated signals designed to improve the detection capability and target resolution through suitable processing at the receiver. In Fig. 8.3(b) the continuous signal is modulated in phase by a *pseudo-random* (PR) code. This causes the signal to appear as broad-band noise, which effectively increases target resolution [15]. The third type of signal carries this process further with the *chirp* or frequency-modulated (FM) signal [3].

As in the passive case the essential tool for echo detection is the correlation of the returned signal with a sampled sequence from the transmitted gated pulse. An early implementation of a correlation receiver used a tapped delay line to carry out the correlation operation. In this analog device a recirculating delay line was used to compress the long duration sonar signal into a short pulse by feeding the output of the delay line back into the input after a delay of NT s, where N is the number of taps of a delay line spaced at intervals of T s. The parallel

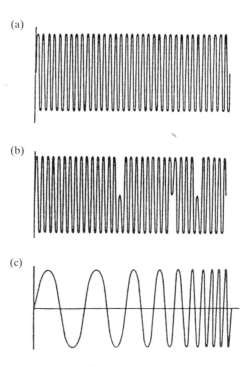

Fig. 8.3 Sonar transmitted signals. (a) Sinusoidal; (b) phase-modulated; (c) frequency-modulated (chirp).

outputs from the taps of the delay line are then multiplied by similar outputs from a delayed refererence signal to produce the correlated output. This is known as the *DELTIC* (delay line time compression) method, used extensively prior to the advent of fast Fourier transform techniques [16].

The transmission of signal and returned echo through the intervening medium modifies the signal received in a number of ways, principally in its attenuation, phase, and Doppler shift. Applying a Fourier transform method, this leads to a complex correlation operation involving the complex envelope representation of the quadrature components and the use of a square law detector. This is expressed in the frequency domain as

$$z_i = \left| \int X(\omega) Y_a(\omega) \, d\omega \right|^2 \tag{8.7}$$

where $X(\omega)$ indicates the Fourier transform of the received complex signal x_i, and $Y_a(\omega)$ is the Fourier transform of the sampled version of the transmitted signal y_i modified by a to include range and Doppler shift values.

A threshold operation is applied to the value of z_i and those values above the threshold value T indicate that a target is present within a given range and Doppler shift [17]. The complete detection procedure is shown in Fig. 8.4. This simple point-target processing is modified in practice due to the inhomogenuous nature of the underwater transmission path. One such modification is to minimize the effects of multi-path transmission and reflection from spurious targets, noted earlier in connection with passive sonar. This is known as *range spreading*. A second is to accommodate the energy reflected by a moving target. This is called *Doppler spreading* and has the effect of widening the frequency range of the received signal.

One method of dealing with these changes in the reflected signal is to combine the outputs of a number of correlator receivers, each designed for a given spread of range and Doppler shift [17, 18]. One technique applies the FFT in the design of a correlator sonar receiver where this is carried out in the frequency domain. This is shown in Fig. 8.5. Here the

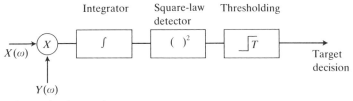

Fig. 8.4 Sonar signal reception.

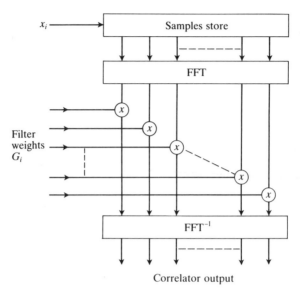

Fig. 8.5 Doppler shift operation.

transform-modify-inverse-transform procedure acts to reproduce a bank of linear filters (Wiener filters), where each set of filter weights G relates to a different range delay. The output consists of a set of correlated signals appropriate to a number of different Doppler shifts. The procedure is followed by a scanning operation, which selects the location of maximum correlation and removes all other contributions.

8.2.4 Sonar imaging

The use of a directional beam of sound waves for the location and identification of underwater objects is applied in a *sonar imaging system*. The object is scanned by the beam to permit a reconstruction of a two-dimensional area of reflected energy rather than simply recording the response from an assumed point source. The process of sonar imaging has close parallels with the formation of scanned images in biomedicine, which we consider in Chapter 9. However since the transmission distance between the energy source and the reflecting object is much longer, different forms of scanning mechanisms are used.

Commonly a linear array of transducers is applied, which act as a generator as well as a detector of acoustic energy. Each element in the array transmits or receives a signal similar to its neighbour, but has a phase difference to it. The principle of constructive interference that applies in the case of such a directional phased array is shown in Fig. 8.6.

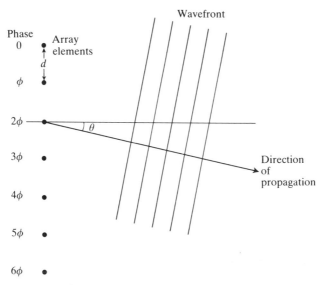

Fig. 8.6 A directional phase array.

At a large distance from the array the radius of curvature of the wavefronts is so large that the waves appear planar. If the phases of the signals driving the individual elements of the array are such that all the elements are driven in phase, then the direction of propagation will be normal to the array. However, if a progressive phase shift is imposed upon the driving signals such that each differs from its neighbour by an amount ϕ degrees, then the wavefronts will add along a line at this angle to the X axis. It can be shown [19] that the angle of divergence θ from the axis is proportional to the phase difference ϕ between adjacent elements. By making slight adjustments to individual phase shift values it is possible not only to direct the beam at a particular angle θ, but also to focus the array on to a focal point in the far field of the array (i.e. at a considerable distance from the array). The electrical phase angle required between a given element and its neighbour is shown to be [19]

$$B = k(r_n^2 + d^2 + 2r_n^2 d \sin \theta)^{1/2} \tag{8.8}$$

where B = the electrical phase angle, r_n = distance from the nth transducer to the focal point; d = inter-element spacing; and θ = required deflection angle.

The interception and detection of the reflected beam is made by a similar (or the same) array of transducers. Again sound incident upon the array from the distant reflector is assumed to be in the form of plane waves. As with phase difference array driving, a relationship exists between the lines of constant phase shown in the diagram and the phase of the detected signal, which changes linearly with distance [20]. Thus the frequency of the array signal will have a relationship with the angle of incidence for the incoming signal. If the square of the Fourier transform for this signal is taken, a direct measure of the energy arriving from a particular direction is obtained and its value may be used in a reconstruction algorithm for deriving the shape of the object.

Two further processing requirements need to be met in a practical sonar scanning device. The assumptions regarding the planar nature of the reflected signals are no longer true in the near field of an array so that, as with a transmitter array, it is necessary to make slight adjustments to the phase of the signal received at each array elements to correct for this. Second, the transformed values actually give *angular information* about the object under examination and it is necessary to convert the polar coordinates into *Cartesian coordinates* so that the actual shape of the object can be displayed correctly. An acoustic imaging system of this type is described in [21] and details of the complex microprocessor control of the beam-forming and processing system are described in [22].

8.3 Seismology

Seismic signals arise from the translation of abrupt earth movements into a continuous waveform by means of a suitable electrical transducer. Due to the mass and elasticity of the earth's crust the waveforms generated will exhibit a multi-degree of freedom damped oscillatory characteristic, represented by the summation of a set of sinusoidal waveforms. The situation is complicated by the fact that this set of waveforms (referred to as *wavelets*) comprising the seismic signal travel at different speeds through a number of different transmission paths to the receiver, each having its own attenuation characteristic. Since the wavelets are essentially sinusoidal, the recorded seismic signal or *seismogram* is amenable to frequency description, and spectral analysis forms a major element in the signal processing procedures.

As with sonar there are two main processing situations, dependent on whether the signal derives from a distant seismic event (earthquake) or from echoes resulting from a locally generated event (explosion). These give rise to a study of the mechanisms of earth movement (*earthquake analysis*) and a study of the structure of the earth's interior (*explosion seismology*). In the first case the seismic signals arise from natural phenomena and are detected some distance from their source. The signals are often quite small, obscured by noise, and subject to multi-path distortion. Explosion signals, on the other hand, travel from a source, usually located close to the earth's surface, and radiate into the earth. The signals received at one or more detection sites are the reflected

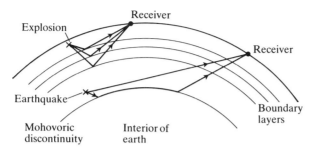

Fig. 8.7 Earthquake and explosion seismology.

signals from the boundaries of layers of sedimentary material comprising the earth's crust. The location and extent of these layers need to be known.

The two cases are shown in simplified diagrammatic form in Fig. 8.7. Note that in earthquake analysis the received signals can arrive from a very distant source. Distant signals are guided towards the receiver along a deep layer boundary known as the *Mohovoric discontinuity,* where the signal suffers little attenuation. A full treatment of the signals and their characteristics is to be found in [23, 24].

8.3.1 Seismic signal analysis

Spectral representation is a major tool for both earthquake analysis and explosion seismology. In addition, the former requires a means of identification of the initial movements of the earth to seismic signals arriving from different directions. This is necessary to enable a differentiation to be made between an explosive and an earthquake source [25].

The magnitude spectrum is obtained through the periodogram or a modelling technique may be used. The latter has the advantage that fewer parameters are needed to represent the signal adequately. This is important in view of the very large amount of seismogram data continually being recorded (more than half a million seismograms are recorded from California alone each year [26]).

The essence of seismic analysis is to separate out (deconvolve) the primary seismic signal from its echo reflections and accompanying noise. Modelling techniques are very effective in doing this. The two most frequently used in seismology are *predictive deconvolution* and *homomorphic deconvolution.* Both recognize that an earthquake signal x_i consists of a series of wavelets w_i, each essentially having the same shape but differing from each other in amplitude and their time of arrival. In predictive deconvolution x_i is considered to be the result of convolving a waveform w_i with a series of random 'spikes' s_i, each spike representing the arrival of a separate reflection of the signal with a given time delay and having an amplitude representing the strength of the reflection at that point, i.e.

$$x_i = w_1 \star s_i. \tag{8.9}$$

Given enough of these spikes spread over the period of the signal they can be regarded as an uncorrelated white noise signal. Hence the autocorrelation of x_i is the same as the autocorrelation of w_i (except for a scale factor which can be ignored). The autocorrelation is used to determine a set of predictive operator coefficients for the case of a minimum delay waveform, assumed to be the basic wavelet for the seismic signal [27]. These predictor operator coefficients c_i are then used as an inverse operator to convolve with the signal x_i to produce a smoothed seismic signal without the multiple reflections.

Seismic analysis by homomorphic processing was discussed earlier in connection with echo detection (Section 8.2.1). It is a powerful technique that does not require *a priori* knowledge of the shape of the seismic wavelet; neither is it necessary to assume a minimum delay for this [28]. Unlike the echo identification process described in Section 8.2.1, which makes use of the power cepstrum, deconvolution in homomorphic filtering requires the complex cepstrum discussed earlier (Section 7.3.8). This is because it is necessary to identify a number of features in the seismic signal extending over a long time period, where the phase of the signal assumes some importance. The complex cepstrum K_c of the seismic wavelet w_i and its accompanying reflections, s_i, tend to occupy disjoint time intervals. Thus if the seismic source has a relatively smooth spectrum then w_i will tend to be concentrated near to its origin. If the reflection series s_i has a minimum phase, then it contributes to the complex spectrum only for a time equal to or greater than the time of the first reflection. This tendency for different components to occupy different time intervals in the pseudo-time cepstrum domain permits the possibility of separating these components by 'low-time filtering' of the cepstrum before inverse transformation is carried out to recover the original wavelet. This is the reason why the complex cepstrum rather than the power cepstrum is needed [29]. The calculation of the complex cepstrum is a major computational task in carrying out homomorphic filtering with discrete data. The FFT is used to obtain equally spaced frequency samples, which are subject to a logarithmic operation and then inversely transformed. It can be shown that since the complex cepstrum is computed from samples of its Fourier transform, a time-aliasing of the complex cepstrum output can result. To avoid this, the phase of the transform must be generated as nearly continuous with frequency as possible. Typically the phase is computed; modulo-2 and a special 'unwrapping' procedure are used to obtain very many more samples of phase values in the Fourier spectrum, amounting to almost a continuous phase curve. An adaptive integration routine for this is described by Tribolet and forms part of a general homomorphic process [30].

8.3.2 Seismic arrays

Earthquake signals are almost always quite small and it is common to use a phased array of seismic detectors, often extending in two or more directions over an area of several square kilometers [31]. Not only does this improve the SNR by a factor of $\sqrt{n}$ or more, where n is the number of elements, but by suitable phasing of the addition of signals from the elements in the array, it is possible to enhance the sensitivity for one particular direction and so determine the direction of arrival for the seismic disturbance. The process was referred to earlier as *beam-forming*.

The principle of beam-forming is shown in Fig. 8.8 for two adjacent array elements A and B. Signals from a single point source at angle θ from the beam axis arrive at A and B with a path length phase difference u

$$u = 2\pi(d/\lambda)\sin\theta \qquad (8.10)$$

where d is the inter-element spacing and λ is the wavelength of the signal.

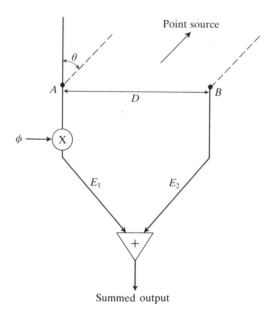

Fig. 8.8 Beam forming with a seismic array.

The signal from B is

$$E_2 = b \cos(\omega t + u) \qquad (8.11)$$

where b is the peak signal amplitude.

If an offset phase ϕ is introduced into the signal at A, giving

$$E_1 = a \cos(\omega t + \phi) \qquad (8.12)$$

then E_1 and E_2 may be combined to obtain a real valued output

$$\mathrm{Re}[E_1 E_2] = [(ab/2)\cos(\omega t + \phi - u)]. \qquad (8.13)$$

This is a constant complex number, the amplitude of which is proportional to the product of the amplitudes of the signals received by the two elements and the phase of which equal to the path length phase difference u. Choice of $\phi = u$ will enable maximum sensitivity to be obtained for the directivity phase angle θ. Similar principles are applied to a large array of n element values.

Processing the summed signal is carried out using the methods described earlier. Continuous recording at the large fixed seismic arrays is often made on a multichannel basis. This enables phasing delays to be inserted in individual channels to suit a particular detection requirement (it is possible for example to include a null value in one particular direction, corresponding to the known direction of a source of seismic noise). Since the amount of data collected in this way is so large, automatic seismic event detection is carried out enabling a selection of data to be obtained, based on some event criterion.

Many detection systems use the amplitude of the signal as the basis for a decision. However, the waveforms for seismic events differ from background noise not only in amplitude but also in frequency content. Hence the interest in using a detector based on a transform of the data, where changes in amplitude and frequency can be sensed and used for determining the occurrence of seismic events. While the FFT can be used in a detection algorithm, it is too slow for many real-time applications. A particularly successful algorithm for automatic event detection has been described by Goforth and Herrin, using the FWT [32]. The seismic signal is digitized in real time, with processing being carried out on a window of 64 samples of data that are updated by 32 samples as each new set of samples is processed, this enables an overlap of detection and processing to occur. This procedure implies, of course, that the processing for event detection must be capable of operating faster than real time. The current set of data is transformed and the Walsh coefficients, X_k weighted in such as way as to *whiten* the spectrum of the noise (i.e. to flatten its spectral characteristics). In some versions of the algorithm the coefficients are further weighted by 0 or 1 to isolate the expected sequency band of the signal. Event detection is obtained by comparing the median value of the samples, which are summed to give an M value of

$$M = \sum_k X_k,$$

with an adapted threshold value. The absolute values of the coefficients, rather than their squares, are summed since this appears to give a more stable parameter in this application and is much faster to compute. The detection threshold is defined by

$$T = V_{50} + K(V_{75} - V_{50}) \qquad (8.14)$$

where T is the threshold, V_{50} the median of the distribution of the previous 512 values of M, V_{75} is 75 per cent of the distribution of the previous values, and K is an arbitrary constant.

If the current value of the sum of the absolute values of X_k, M exceeds the threshold T, a signal is called. If it does not exceed the threshold, the sum of the absolute values is ranked among the previous set of values, the oldest value being discarded. In practical terms this adaptive threshold means that the threshold follows a running average of the signal with a time lag. If the signal is increasing, then an increase in threshold level occurs and an event is indicated if the signal input continues to rise. If the signal decreases the average value follows it down so that small insignificant events can still be detected. The algorithm has been implemented at several sites in the United States, and in Norway on the NORSAR seismic array [33].

8.3.3 Seismic data compression

Another area where the Walsh transform has proved useful is in the reduction of the volume of seismic data records for storage or onward transmission. A simple data compression scheme involves a Walsh transformation followed by sequency band-limiting to reduce the information transmitted to the first spectral lobe of the transformed signal (Fig. 8.9). Inverse transformation and frequency bandpass filtering restores the compressed seismic signal. A data compression of about 7:1 is achieved by Wood without too severe a signal degradation [34]. It has been shown however that when applied to pattern recognition for seismic signals, e.g. discrimination of an earthquake signal from an explosion source, both the DCT and the FWT are less effective than the FFT [35]. The reason appears to be that the cosine and Walsh transforms spread the signal energy to higher spectral components, a factor which has also been noted by Kennett [36], and band-limiting removes this component from the signal, thus distorting the recovered signal.

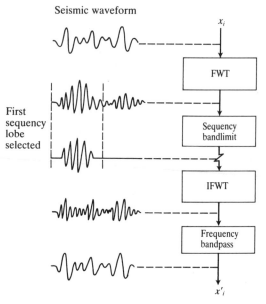

Fig. 8.9 Seismic data compression.

8.4 Radar

Echo detection through transmitted radar pulses is similar to active sonar, but it uses much higher frequencies and operates in a more favourable environment. The returned signals are however considerably smaller and the SNR is generally worse than that experienced in sonar, placing the emphasis on efficient detection methods.

The radar transmitter provides a periodic sequence of pulsed sinusoidal waves generated at microwave frequencies and concentrated by means of a directional aerial as a narrow beam of energy in the direction of the expected target. Some portion of this energy is reflected and detected (usually by the same aerial) at a sensitive receiver after a finite time, t s, following the transmitted pulse. The time taken for the transmitted pulse to travel a distance of r metres to the target and to be reflected and returned to the aerial is

$$t = 2r/c \tag{8.15}$$

where c is the velocity of light $(3 \times 10^8 \text{ m s}^{-1})$.

If T is the pulse repetition interval, the maximum unambiguous range is given by

$$r_{max} = cT/2. \tag{8.16}$$

Other information about target location (azimuth and elevation angle) can be obtained from the known direction in which the aerial is pointing at the precise time of transmission and reception.

To maximize the SNR the transmitted bandwidth is made a reciprocal of the pulse width and as the desired range and resolution are extended there are obvious difficulties in providing the peak power required in the short time available. A way out of this difficulty is to transmit a longer pulse than that finally required and to code this pulse in time in such a way as to enable compression of the reflected pulse energy into a shorter pulse on reception. The most common coding used is a linearly swept frequency, the *chirp signal* that we met earlier in connection with CCD devices (Section 6.4.2). Modern digital radars also make use of a binary code having precise autocorrelation characteristics. The most well known of these is the *Barker code* [37, 38], in which the autocorrelation function of the received waveform consists of a large central spike with an equal number of uniformly low smaller spikes on either side. This results in a significant improvement in the inherent range resolution over a single wide rectangular pulse.

8.4.1 *Moving target indication*

In many practical radar systems both range and velocity of a moving target are required. This gives rise to a *moving target indication* (MTI) radar which extracts information about the moving target from a noisy (or clutter) background. In MTI radar the Doppler shift in frequency is used to discern moving targets which may produce signals considerably smaller in magnitude than those received from nearby fixed targets [39].

The principle is to detect the difference in frequency between the returned echos and a reference frequency derived from the frequency of the transmitted carrier pulse. Defining this difference frequency as f_d, with f_0 as the carrier frequency, then the target velocity V can be obtained using the *Doppler equation*

$$V = cf_d/2f_0. \tag{8.17}$$

A problem however lies in distinguishing between these echoes and the clutter signals, which may be several orders of magnitude larger than the desired signal. Fortunately this clutter is concentrated near zero frequency and by employing Doppler frequency sensitive filters it is possible to remove most of the clutter. In Doppler radar a periodic set of pulses is transmitted, known as the *burst waveform*. The Doppler spectrum of such a burst is periodic, with a period equal to the pulse repetition frequency of the burst. Assuming that the clutter is at zero frequency, then for a given range this can be removed by subtracting the output from two successive pulses spaced d s apart. A simple filter to carry this out is called a *two-pulse canceller* (Fig. 8.10). This is actually an

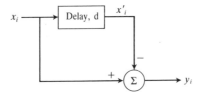

Fig. 8.10 A two-pulse canceller.

elementary FIR filter (see Section 1.3.1). In a complete design for an MTI radar the Doppler spectrum of the clutter is assumed to be located in frequency bands disjoint from the band occupied by the moving target signals. An FIR high-pass filter is then followed by a set of narrow band filters tuned to the specific Doppler frequencies of interest. Examples of the design of MTI processors are described in [40] and [41].

8.4.2   Transform echo-detection methods

The most elementary detecting scheme is to subject the received pulse, illustrated in Fig. 8.11(a), to thresholding, whereby a signal crossing the threshold T_1 indicates the presence of target information and the delay corresponding to this crossing is a measure of the range. In cases of high noise level such as that shown in Fig. 8.11(b), this would result in a high proportion of false alarm readings. However by considering the *similarity* between the received pulse compared with the transmitted pulse, a different criterion may be adopted that is not dependent on pulse

(a)

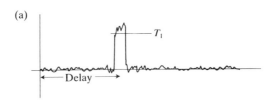

(b)

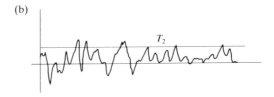

Fig. 8.11 Radar thresholding operation. (a) Low-noise level; (b) high noise level.

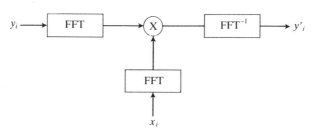

Fig. 8.12 Matched filtering.

amplitude. As in passive sonar cross correlation between the received signal and the transmitted waveform can achieve this. This is the *matched filter* method shown in Fig. 8.12; it is a special case of the Wiener filtering considered earlier. A large value $R_{xy}(.)$ indicates a high similarity between x_i and y_i and the location of a correlation peak along the time axis is a measure of the delay at which the similarity occurs. For reasons of computational economy the correlation is carried out through the frequency domain by the use of the FFT. The implementation of the matched filter using this method requires $N(1 + 4 \log_2 N)$ multiplications, assuming that the matching sequence x_i is stored as a precomputed value.

Since in principle the matched filter (Wiener filter) can operate with any orthogonal transform, attempts have been made to use the faster sequency transforms in order to reduce the number of multiplications required. With the generalized matched filter shown in Fig. 8.13, Polge [42] has shown that by suitable selection of the weighting sequence G_i, a maximum SNR can be achieved for orthogonal transforms such as the FWT.

Using the FFT the values for G_i are simply the Fourier-transformed coefficients of x_i. With the FWT however, a difficulty arises due to the fact that the Walsh transform is not invariant with phase shift and this requires that a different weighting sequence G_i be used with each different delay value. While this can be implemented through the use of a bank of Wiener filters it is uneconomic and an alternative method, suggested by Polge, is to apply dyadic correlation calculated at a fixed set of delays, where the dyadic and arithmetic correlation are shown to be

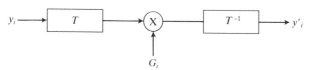

Fig. 8.13 Generalized matched filtering.

equal [43]. To achieve this it is necessary to sample the transmitted pulse at K discrete values, where $K = 2^k$ and $K \ll N$. The dyadic and arithmetic correlations are then equal at particular values of i, namely, $i = 0, K, 2K, \ldots$, and from these dyadic correlations all the samples of the cross-correlation can be derived. This reduces the number of multiplications required to $K . N$, which will be smaller than $N(1 + 4 \log_2 N)$ when the pulse width is smaller than the range window, N.

8.4.3 Other methods

Conventional Doppler processing uses filtering and Fourier transformation. In order to extract the target signal from background clutter a relatively long sample of the signal is required. This brings about difficulties in using the high search rates required in modern radar systems. A method based on *maximum entropy* (MEM) spectral analysis has been effective in overcoming this problem in some circumstances [44]. As described in Section 1.5.4 a set of adaptive filter coefficients are determined from the received signal and applied to the MEM. The logarithm of the MEM spectrum is estimated and compared with a threshold value to determine whether a target is present or not. It has been shown that where the input clutter spectrum has a narrow bandwidth then the MEM Doppler processor is significantly better than the conventional Fourier method [44]. This advantage is not maintained for all types of clutter signal however, although it remains an effective method [45].

8.4.4 Non-sinusoidal radar

Recent interest in the radiation and detection of non-sinusoidal signals has resulted in the development of a number of specialized radar systems offering some advantages over conventional designs. The principle reason for this lies in the problems encountered with the use of sinusoidal carriers as the transmission frequencies extend into the GHz region. The transmission problems are most clearly recognized in radar applications. A major difference between the use of sinusoidal and non-sinusoidal carriers is, of course, the bandwidth occupied by the transmission. This has been defined in terms of a *relative bandwidth* as a ratio of bandwidth to carrier frequency. A more general definition is given by Harmuth [46] as

$$\eta = (f_h - f_j)/(f_h + f_j) \qquad (8.19)$$

where f_h is the highest and f_j is the lowest frequency of interest. Typical radio signals used for communications or radar have relative bandwidths of the order of $\eta = 0.01$ or less, whereas most of the signals we desire to transmit, for example, sound and vision signals, have η values close to or equal to unity. A typical solution for transmission purposes is to transform these large relative bandwidths into smaller bandwidths using carrier modulation. A natural limitation is placed on this process of increasing the carrier frequency to achieve a small relative bandwidth by the attenuation effects of rain, fog, and molecular absorption. This limit is reached when the carrier frequency exceeds about 10 GHz [47]. Thus if radar transmission are limited to this maximum carrier frequency, then the resolution is also limited. Channel capacity for communications is subject to similar limitations [48].

Early attempts to apply sequency methods to radar systems considered the consequences of replacing a sinusoidal wavetrain, modulated or unmodulated, by a Walsh carrier. Lackey [49] observed that the resolution of point targets can be enhanced if the target area is illuminated with a Walsh wave rather than a sinusoidal signal. He gives an example to show that the difference between a reflection from two separate targets in terms of their summation is no longer a small perturbation of the reflected signal but a major change in the summed and reflected signal, leading to an increase in resolution for the Walsh case. Other approaches to the utilization of Walsh waves in radar include synthesis methods where a number Walsh functions are added to provide a modulating waveform having optimum characteristics. Some analytical work has been carried out by Rihaczek, which showed that the choice of optimum radar waveform is greatly dependent on the exact nature of the target and type of discrimination required [50]. The Walsh sequence has some advantages in this respect. The problem is also discussed by Griffiths and Jacobson [51].

While most of these early investigations looked into the possibilities of using complete Walsh or Rademacher series, more recent developments have considered instead the transmission of very narrow pulses by utilizing a frequency band extending from 1 to 10 GHz. This has become known as *impulse radar* or *carrier-free radar* and this chapter concludes with a brief look at one extremely successful application of this new technique. This is termed *into-the-ground impulse radar,* first proposed by Cook in 1960 [52], primarily to detect buried clay pipes and other artifacts, difficult to detect using other methods. Inhomogeneities in the ground make the attenuation of electromagnetic waves increase with frequency very rapidly, just as raindrops an fog droplets in air inhibit high-frequency radar transmission. In order to provide an acceptable resolution, such into-the-ground radars must use pulses with a duration of about 1 ns. For reasons given earlier this is not a practicable proposition with carrier-borne techniques. The operation of this type of radar is illustrated in Fig. 8.14 (included by courtesy of Professor H. Harmuth and Academic Press). The radar device is mounted such that it can

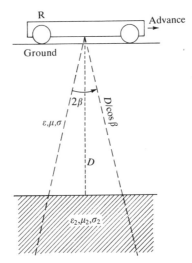

Fig. 8.14   Into-the-ground radar. (From Harmuth [46].)

traverse the surface of the ground. A pulse is radiated into the ground at intervals of the order of microseconds to milliseconds. A layer at the depth D with a discontinuity of the dielectric constant ε, magnetic permeability μ, or in the conductivity σ, will reflect the signal, which is detected after a delay appropriate to the depth of penetration and reflection back to the surface [53]. The range of detection extends over $2D \tan \beta$ at level D, where the beam angle of the radar equals 2β. The discontinuity will be seen in the record of the reflected signal amplitude as a band of width $[D/\cos(\beta - D)]$ following a delay proportional to the depth D. A number of examples of actual recording are given by Harmuth in [46], and such devices are now widely used for subsurface investigations related to geotechnical, engineering, military, or archaeological applications [54, 55].

References

1. Stephens, R. W. B., ed. *The sea as an acoustic medium in underwater acoustics* pp. 1–22. Wiley, London (1970).
2. Ulrick, R. J. *Principles of underwater sound for engineers.* McGraw-Hill, New York (1967).
3. Van Trees, H. L., *Detection, estimation and modulation theory.* Wiley, New York (1968).
4. Griffiths, J. W. R., Stocklin, P. L., and Schooneveld, C. Van, ed. *Signal processing.* Academic Press, London (1973).
5. Tacconi, G., ed. *Aspects of signal processing with emphasis on underwater acoustics.* D. Reidel, Dordrecht (1977).
6. Thomes, R. S., Moldon, J. C., and Ross, J. M. Shallow water acoustics

related to signal processing. in *Signal processing* (ed. J. W. R. Griffiths, P. L. Stocklin, and C. Van Schooneveld) pp. 281–98. Academic Press, London (1973).

7. Carter, G. C., Knapp, C. H. and Nuttall, A. H. Estimation of the magnitude-squared coherence function via overlapped fast Fourier transform processing. *IEEE Trans. Audio Electroacoust.* **AU-21** (4), 337–44 (1973).

8. Carter, G. C. and Knapp, C. H. Coherence and its estimation via the partitioned modified chirp-Z transform. *IEEE Trans. Acoust. Speech Signal Process.* **ASSP-23** (3), 257–64 (1975).

9. Hassab, J. C. Passive tracking of a moving source by a single observer in shallow water. *J. Sound. Vib.* **44** (1), 127–45 (1976).

10. Carter, G. C. Methods for passive locating an acoustic source. IEEE Conference on Acoustics, Speech and Signal Processing, Piscataway, New Jersey (1977).

11. Chen, C. H. Digital processing of marine seismic data, IEEE Proceedings of an International Conference on Engineering in the Ocean Environment, Piscataway, New Jersey, pp. 346–350 (1972).

12. Hassab, J. C. and Boucher, R. A probabilistic analysis of time delay extraction by the cepstrum in stationary Gaussian noise. *IEEE Trans. inf. Theory* **IT-22**, (4), 444–54 (1976).

13. Hassab, J. C. and Boucher, R. Analysis of signal extraction, echo detection and removal by complex cepstrum in the presence of distortion and noise. *J. Sound. Vib.* **40**, 321–35 (1975).

14. Hahn, W. R. Optimum signal processing for passive sonar range and bearing estimation. *J. acoust. Soc. Am.* **58** (1), 201–7, (1975).

15. Kennedy, R. S. and Lebow, J. L. Signal design for dispersive channels. *IEEE Spectrum* **1**, 231–237 (1964).

16. Allen, W. B. and Westerfield, E. C. Digital compressed time correlators and matched filters for active sonar. *J. acoust. Soc. Am.* **36**, 121–39 (1964).

17. Oppenheim, A. V., (ed. *Applications of digital signal processing.* Prentice-Hall, Englewood Cliffs (1978).

18. Kennedy, R. S. *Fading dispersive communication channels.* Wiley, New York (1969).

19. Robinson, G. P. S. Acoustic imaging. Ph.D thesis, University of Newcastle-upon-Tyne, pp. 39–61 (1978).

20. Keating, P. N., Koppelman, R. F., and Mueller, R. K. Complex on-axis holograms and reconstruction without conjugate images. In *Acoustic holography* pp. 515–26. Plenum Press, New York (1974).

21. Yuen, C. K., Beauchamp, K. G., and Robinson, G. P. S. *Microprocessor systems and their application to signal processing.* Academic Press, London (1982).

22. Kennair, J. L., Posso, S. M., and Taylor, B. Microprocessor aids to acoustic imaging. IEE Colloquium on microprocessors in the Marine Industries, IEE, London (1982).

23. Claerbout, J. F. *Fundamentals of geophysical data processing with applications to petroleum prospecting.* McGraw-Hill, New York (1976).

24. Bolt, B. A. *Nuclear explosion and earthquakes: the parted veil.* Elsevier, Amsterdam (1978).

25. Chen, C. H. *Computer-aided seismic analysis and discrimination.* Elsevier, Amsterdam (1978).

26. Anderson, K. R. Automatic processing of local earthquake date. Ph.D. thesis, Massachussets Institute of Technology, Cambridge, MA (1978).
27. Peacock, K. L. and Treitel, S. Predictive deconvolution—theory and practice. *Geophysics* **34**, 155–69 (1969).
28. Ulrych, T. Applications of homomorphic deconvolution to seismology. *Geophysics* **36** (4), 650–60 (1971).
29. Tribolet, J. M. Application of short-time homomorphic signal analysis to seismic wavelet estimation. In *Computer-aided seismic analysis and discrimination* (ed. C. H. Chen). Elsevier, Amsterdam. pp. 75–96 (1978).
30. Tribolet, J. M. *Seismic applications of homomorphic signal processing*. Prentice-Hall, Englewood Cliffs (1979).
31. Green, P. E., Frosch, R. A., and Romney, C. F. Principles of an experimental large aperture seismic array (LASA). *IEEE Proc.* **53** (12), 1821–33 (1965).
32. Goforth, T. and Herrin, E. An automatic seismic signal detection algorithm based on the Walsh transform. *Bull. seism. Soc. Am.* **71**, 1351–60 (1981).
33. Veith, K. F. Seismic signal detection algorithm. NTIS Report No. AD-A110 186/4. Teledyne Geotech, Garland, Washington, DC, (1981).
34. Wood, L. C. Seismic data compression methods. *Geophysics* **39**, 499–525 (1974).
35. Chen, C. H. Comparison of orthogonal transforms for teleseismic data. NTSI Report No. AD-A 006232. Teledyne Geotech, Garland, Washington DC (1974).
36. Kennett, B. L. N. Short-term spectral analysis and sequency filtering of seismic data. In *Exploitation of seismograph networks* (ed. K. G. Beauchamp) pp. 283–96. Noordhoff, Leiden (1975).
37. Woodward, P. M. *Probability and information theory with application to radar*. Pergamon Press, New York (1955).
38. Rihaczek, A. W. *Principles of high resolution radar*. McGraw-Hill, New York (1969).
39. Barton, B. K. *Radar system analysis*. Prentice-Hall, Englewood Cliffs, (1964).
40. Delong, D. F. and Hofsletter, E. M. On the design of optimum radar waveforms for clutter rejection. *IEEE Trans. inf. Theory* **IT-13**, 454–63 (1967).
41. Muehe, C. E. Cartletge, L., Drury, W. H., Hofstetter, E. M., Labitt, M., McCorisan, P., and Sferriro, V. J. New techniques applied to air traffic control radar. *IEEE Proc.* **62** (6), 716–723 (1974).
42. Polge, R. J. and Bhagavan. B. K. Signal processing techniques for real-time radar applications. UAH Research Report 144, 1. University of Alabama, Huntsville (1973).
43. Polge, R. J. and Bhagavan, B. K. Investigation of transform techniques for digital processing of radar signals. NTIS Washington, AD/A-003/2ST (1974).
44. Haykin, S. and Chen, H. C. The maximum-entropy spectral estimation used as a radar Doppler processor. Proceedings of the RADC Spectral Estimation Workshop, Rome Air Development Center, Rome, NY. (1979).
45. King, W. R. Applications for MESA and the prediction error filter. Proceedings of the RADC Spectral Estimation Workshop, Rome Air Development Center, Rome, NY (1979).
46. Harmuth, H. F. *Non-sinusoidal waves for radar and radio communication*. Academic Press, New York (1981).

47. Harmuth, H. F. Fundamental limits for radio signals with large relative bandwidth. *IEEE Trans. electromagn. Compat.* **EMC-23,** 37–43 (1981).
48. Harmuth, H. F. Interference caused by additional radio channels using non-sinusoidal carriers. IEEE Electromagnetic Compatibility, Zurich, Switzerland, (1977).
49. Lackey, R. B. The wonderful world of Walsh functions. Proceedings of a Symposium on Applications of Walsh Functions, Washington DC, AD744650, pp. 2–7 (1972).
50. Rihaczek, A. W. Radar waveform selection: a simplified approach. *IEEE Trans. Aerosp. electron. Syst.* **AES-7,** 6 (1971).
51. Griffiths, L. J. and Jacobson, L. A. The use of Walsh functions in the design of optimum radar waveforms. In *Proceedings of a symposium on the Applications of Walsh functions* (ed. H. Screiber and G. P. Sandy). IEEE Publication No. 74CH0861-5EMC, New York (1974).
52. Cook, J. C. Proposed monocycle pulse very high frequency radar for airborne ice and snow measurement. *AIEE Trans. Commun. Electron.* **79,** 588–94 (1960).
53. Chapman, J. C. Experimental results with a Walsh wave radiator. *Nat. Telecom. Conf. Rec.* Vol. III, 44.2.1–44.2.3 (1976).
54. Campbell, K. J. and Orange, A. S. A continuous profile of sea ice and freshwater ice thickness by impulse radar, *Polar Rec.* **17** (106), 31–41, (1974).
55. Ulrikson, C. P. Application of impulse radar to civil engineering. *Geophysical Survey Systems Inc. Report,* Hudson NH (1982).

9

BIOMEDICINE

9.1 Introduction

The two major divisions of biomedical signal processing are:

(a) biological signal processing involving *electrocardiographic* (ECG), *electroencephalographic* (EEG), and biorhythm data;

(b) diagnostic medical signal processing, principally body scanning or *tomography*.

This division also happens to be a dimensional one where (a) is concerned generally with *single-dimensional* (1-D) processing and (b) with *multi-dimensional* or 2-D image processing. In tomography the 2-D image obtained is displayed as a 'slice' taken from a three-dimensional model; occasionally genuine 3-D data is obtained [1, 2]. A common problem running through all of these areas of application is the imprecision of the medical characteristics sought for. Usually in signal processing the characteristics of the signal are well defined and the purpose of the processing is to enhance these or to extract the signal from a noise or irrelevant background. In medical processing the problem is often the identification of an unusual feature within a known area, e.g. a cancerous growth or an internal injury, and the classical 'shape' identification techniques such as the use of templates in matching are difficult to apply. Apart from SNR enhancement—an essential feature of much biomedical processing, where layers of tissue invariably intervene between the source and the signal transducer—processing will be concerned with classification of the received signal. This is a major and specialized subject area and lies outside the scope of this book. Indeed the range of signal processing techniques applied to biomedical processing is itself quite wide and only a brief outline of some of the more important procedures, principally related to domain transformation, is possible in this chapter.

9.2 1-D processing

Single-dimensional medical data processing is concerned usually with continuous processes taking place within the body that can be readily monitored. Blood flow, muscular rhythms, neurological processes, respiration, and heart movement are all of interest to the clinician, and electrical signals having a direct relationship to these activities are relatively easy to obtain and store. In addition to specific diagnostic measurements, there is now a requirement for continuous or on-line patient monitoring. In many of these activities there is a need for patient record data storage or transmission, and bandwidth compression using transform techniques is important here [3, 4]. Transformation is also applied in the analysis of ECG and EEG data. In addition statistical methods are used [5], and there is a growing interest in the superior resolution that can be obtained using auto-regressive analysis [6]. However, these latter topics will not be discussed further here.

9.2.1   ECG analysis

The electrocardiogram gives a measure of the electrical activity of the heart in terms of a continuous time history. The most important characteristic of a normal cardiac cycle is the segment shown in Fig. 9.1, which repeats itself with minor variations once per cycle. This is known as the *QRS cycle,* and corresponds to the electrical activity of the ventricles during a single heartbeat. Variations in the shape and repetition rate of the QRS waveform are important in determining the onset of ventricular fibrillation, so that on-line analysis and classification of this segment of the cardiac cycle forms a vital feature of automatic patient monitoring systems. The three waveform characteristics of the cycle, shown at Q, R, and S, relate to particular activities of the ventricle and each has a characteristic shape that can be recognized by a cardiologist. The QRS cycle is preceded and followed by smaller activity peaks, P and T, which relate to atrial activity and are not as readily recognized, particularly when obscured by random noise. For a number of reasons it is desirable to carry out this pattern recognition process automatically, and methods based on transformation instead of time domain correlation have been extensively used. This has a number of advantages for ECG analysis, including the following:

  1. The bandwidth compression that can be obtained in this way permits economic line transmission of the data and subsequently allows

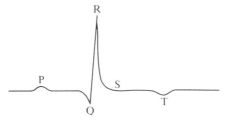

Fig. 9.1   The QRS cycle of cardiac activity.

many more patient records to be stored within a given digital storage medium.

2. Faster identification can be achieved by using fewer significant parameters. This is valuable for on-line patient monitoring systems.

3. Cases of function abnormality are more easily recognized, since they are often accompanied by enhanced higher-frequency content.

The bandwidth criterion is an important one since information on the patient's heart activity is often obtained at the bedside and may be transmitted over a public data transmission or telephone network to the hospital concerned. Standards for this have been proposed by the American Heart Association and are beginning to be applied in commercial terms [7]. Use of these would impose several problems in transmitting data in uncompressed form over voice-grade telephone lines [8].

A number of orthogonal transformations have been used or proposed for bandwidth compression of ECG data [9, 10]. The operation of an orthogonal transformation data compression algorithm is shown in Fig. 9.2. x_i is the digital representation of the ECG waveform consisting of N data samples of M-bit binary numbers. The compression G produces a transmission sequence Y_i having N' data samples of M'-bit binary numbers where $N' < N$, and/or $M' < M$. To be effective, the synthesis operation at the receiving end must produce an acceptable reconstruction z_i that is close to the original sampled signal x_i, i.e.

$$z_i = x_i + e_i \tag{9.1}$$

where e_i represents an error sequence which is made small. The efficiency of a compression system is determined by:

1. The degree to which z_i approximates x_i;

2. The value of the data compression ratio $d = MN/M'N'$ (eqn (7.18));

3. The complexity of the compression/synthesis processes, shown as G and H in Fig. 9.2.

With ECG signals it is necessary to preserve signal accuracy at the beginning of the QRS cycle, since small errors here can have greater diagnostic significance than in other portions of the cycle. This makes

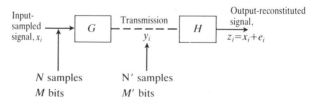

Fig. 9.2 Orthogonal transformation data compression.

orthogonal transform compression methods of especial significance because with these methods the MSE of the reconstructed signal is well-correlated to the maximum difference that can be expected at any point. The normalized MSE is generally chosen therefore as a suitable measure of efficiency for the process, i.e.

$$\text{MSE} = \left[\sum_{i=1}^{N} (x_i - z_i)^2 \right] \bigg/ \left[\sum_{i=1}^{N} (x_i^2) \right]. \tag{9.2}$$

While the Karhunan–Loève transform (KLT) can give optimum results, the cosine transform (DCT), as discussed earlier, can give very similar performance levels with the advantage of having a faster transformation algorithm [11].

The technique for compression is a simple threshold one of discarding all transformed coefficients of higher frequency than N/R, where N is the total set of transformed values and $R:1$ is the *data compression ratio* (DCR) required. A DCR of about $4:1$ can be achieved with a MSE for the DCT quite close to that of the KLT.

Due to the field operational requirements for ECG acquisition and transmission equipment, (the analysis can be carried out elsewhere as mentioned previously), the complexity of the system used must be kept to a minimum. Both the Haar and Walsh transforms have been applied in order to reduce the complexity of the process and to enable real-time operation to be achieved using microprocessor implementation. The Haar function is somewhat similar in shape to the QRS cycle, and using this it is found that only a small number of coefficients are necessary to approximate the waveform to a small MSE (see Section 2.4). In an application of the Haar transform for ECG data compression, Hamba and Tachibara have shown that only 60 coefficients need to be retained out of a total of 128 coefficients to sufficiently identify the waveform [12]. This agrees with the earlier results of Ahmed and Rao, who also compare this result with the performance of the KLT and the DCT [9]. Kuklinski has also carried out a study of the FWT, varying not only the number of sequency coefficients retained but also the number of bits per coefficient [13]. This study shows that a relationship exists between the error of the reconstructed ECG and the values of N' and M' chosen. Specifically, for any desired MSE there exists a unique combination of bits per coefficient M' and fraction of the Walsh spectrum retained N' that produces maximum data compression.

The analysis of the transmitted ECG data can be carried out from either the reconstructed time series or directly from the frequency or sequency domain. In the latter case identification of normal and abnormal cardiac conditions may be made by comparison of the compressed spectral characteristics with previously obtained spectral values from measured and known cases. By using transformation values rather than a sampled time series, considerably fewer characteristics are needed for this process of comparison. Methods of analysis and some results of these tests carried out with the FFT spectrum over various parts of the QRS cycle are given by Cain and Ambosh [14]. This work describes the significant differences in the frequency content of the QRS waveform, which can be observed between normal and abnormal patients, thus permitting medical classification to take place. Analysis methods involving the use of Walsh, Haar, and R transforms are discussed in [5]. The methods described above are essentially pattern matching techniques. A number of statistical transformation techniques for recognition of normal and abnormal ECG waveforms using sequency functions are given by Meffert *et al.*, together with details of algorithms for SNR improvement [16].

9.2.2 EEG analysis

Electroencephalographic signals derive from electrical activity in the cortex region of the brain and are obtained through surface-mounted disc electrodes affixed to the surface of the scalp. The signals are small and associated with unwanted noise, so that amplification and filtering form an essential preliminary to the signal processing. Processing of EEG signals is based principally on the development of time series models for the process using spectral analysis to extract parameters of significance. The information abstracted is used in the diagnosis of neural disorders and to enhance the understanding of neural processes. Considerable clinical interest is paid to the identification of dominant rhythms within the complex EEG signal, particularly since it has been demonstrated by several investigators that the harmonic component of pathological EEGs is far larger than that of normal EEGs [17]. These spectral patterns can be recognized easily without the extensive training required to interpret the original time domain EEG recordings. For this reason domain transformation plays a significant part in EEG analysis at a fairly low analysis level.

The early development of a frequency analyser for EEG interpretation by Baldock and Walter [18] established a pattern for spectral analysis that has continued with the later digital spectral analysis through Dumermuth, Fluhler, and others in which the FFT is used [19, 20].

Since the EEG signal is a non-stationary one, some care is needed to extract accurate mean frequency values. A common procedure, noted in earlier chapters, is to carry out consecutive analysis on a number of time segments, where the time shift from segment to segment is in the order of seconds and the segments themselves are essentially statistically independent of each other [21]. The periodogram of each segment is taken and ensemble averaging of the spectra from each segment is carried out to achieve a final estimate for the entire signal interval. This is shown to give a consistent estimation for the spectrum [22].

EEG signals are essentially sinusoidal in character, so that frequency analysis using the FFT is a natural choice. A number of advantages have been given for using a sequency approach with the Walsh or Haar transform. These have been principally related to the design simplicity of the data acquisition and analysis equipment, which needs to be portable for clinical reasons [23–25]. In recent years however these advantages have become eroded by the availability of cheaper FFT integrated circuits and the increasing power of the microprocessor, so that the comparison between the FFT and FWT is now more properly assessed on a performance basis. Recent work by Jansen has shown that the FWT-derived traditional spectral features are not as useful as the FFT in the classification of short EEG segments, and that difficulties are caused by the appearance of significant secondary peaks in the FWT spectrum that are not present in the FFT spectra and do not reflect visually observable rhythms in the actual time series [26].

The power spectrum for the EEG signal can also be estimated through statistical means and some work on the use of *autoregressive moving average* analysis (ARMA) is described by Ahlbom and Zetterberg [27]. This approach has proved of value in non-stationary analysis and is thus very applicable to the EEG problem.

The output of spectrum analysis described so far gives no substantial data reduction. The large mass of spectral data produced needs to be reduced considerably if comparative studies are to be carried out on a collection of clinical records. One method being pursued is that of *parameter extraction,* somewhat similar to formant analysis in speech (Chapter 7). The spectrum is searched and values are recorded for intensity, frequency bandwidth, and area of significant peaks. Peak statistics can give the mean and standard deviations of peak frequencies as a factor of linear peak power. The bandwidth is determined as the percentage of the area above the virtual peak base, i.e. from a line connecting the troughs in the frequency spectrum on either side of a peak, relative to the total power. Statistical analysis of the parameter database so formed is the subject of much consideration between the clinician and analyser [28] and, as in other areas of biomedicine data analysis, is very dependent on experience of processed data characteristics in relation to clinical conditions.

The most recent developments in EEG data compression and analysis use the methods of *syntactive description.* A typical EEG signal of four-channel data is shown in Fig. 9.3. Each 100 s of the waveform is divided into a number of equal-length segments of perhaps 1 s duration. The smoothed spectral density characteristics for each segment are taken

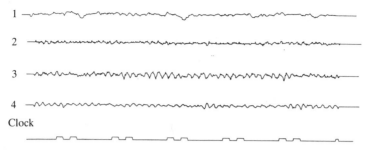

Fig. 9.3   A four-channel EEG signal.

for the four channels and classified into a limited number of spectral shapes, considered as representing typical features of an average EEG signal. With a spectral description limited to say 10 features there would be a total of 400 separate features selected to classify the four-channel EEG waveform out of a possible total of 4000 such features. A very considerable reduction in the amount of data required to represent the complete EEG waveform is thus achieved. Further, by considering each identified feature as an element in a linguistic grammar—rather like consecutive vowel sounds in speech—it is possible to define an algorithm that can establish links between neighbouring features and so enable hierarchical patterns of features for the complete waveform to be evolved. These have been called *parse trees,* and experience has shown that their shape and constituent features can provide useful diagnostic indications to clinicians. This type of analysis is essentially statistical in form however, and apart from a brief mention in the next chapter in the context of pattern recognition, it will not be considered further.

9.3   2-D processing

Two-dimensional images of parts of the body generally obscured by layers of tissue are vital for medical diagnosis. The value of conventional X-ray photographs have been appreciated in the medical field since their discovery by Röntgen in 1895. Recent work in combining ordinary X-rays with sophisticated computer signal processing to generate a display of body tissues unencumbered with the shadows of other organs has caused a revolution in the medical field. It has been responsible for the award of two Nobel prizes in medicine and has been applied successfully to several other applications in the physical sciences [29]. The technique used is a reconstruction method in which many different X-rays of the subject are taken at various angles and the results are digitized and stored in a database. This is subsequently processed to extract a reconstructed image of a cross-section through the three-dimensional subject [30]. Similar principles are used in *emission-computed tomography* (ECT) [31], *nuclear magnetic resonance imaging* (NMR) [32], and in *ultrasonic imaging* (USI) [33].

All of these techniques are referred to as *computer-assisted tomography* (CT) [34–37], in which the main function of the computer processing is to produce a reconstruction of some characteristic (usually tissue density) of a slice taken through the subject being investigated. A fundamental characteristic of a CT image is that the reconstructed value at any spatial point is (ideally) uncontaminated by values of the true density outside the neighborhood of the point. This is referred to as a 'clean image'. A clean image is impossible with conventional optics because wave motions are continuous in space and time. A wave motion brought to a focus necessarily converges before and diverges after the point and thereby unavoidably contaminates the image observed by density values within the double-cone of the focusing region.

9.3.1 Radioisotope scanning

A less sophisticated but also valuable method of medical diagnosis is the two-dimensional scanning of radioactive nuclides that are localized in or surrounding the object under study. This is known as *radioisotope scanning* and is generally carried out on a line-by-line basis by linear movement of a suitable radio-detecting device across the subject area [38]. This form of scanning, like conventional X-ray imaging, produces a 2-D image without the need for image angular reconstruction. A mapping of the internal organ is obtained by determining the spatial distribution of the radioactive nuclides through detection of gamma rays emitted under a continuous process of decay. The method most frequently employed is that of rectilinear scanning [38, 39], using a photoscintillation detector (the process is often termed *scintiography*) to scan the selected area with a television-like raster movement. Some processing of the data is required to overcome the poor resolution of the imaging system and to suppress the noise inherent in both the distributed nature of the detector and the statistical nature of radioactive decomposition. Such image enhancement can be carried out on the complete stored image after this has been built up line-by-line from the digitized output of the scanning detector.

An alternative that is simpler and more compatible with the analog detected data is to use a 1-D line-by-line processor and to process the data in real time [40]. A series of filtering and smoothing operations are carried out on each line of the data before this is stored and displayed. A Wiener filtering method is frequently used involving a transformation, coefficient weighting, and inverse transformation (see Chapter 3) [41]. Selection of coefficient weights are discussed in [38, 40], and acceptable results can be obtained through a combination of a normal Gaussian filter and a low-pass filtering function.

9.4 Computer-assisted Tomography

9.4.1 X-rays

A conventional X-ray image covers an area in a single view. With computer-assisted tomography the area is scanned as a single line and the process is repeated at different angles to the subject over the entire subject area. Applying parallel geometry calculations to this data, the resulting CT image has the form of a cross-section taken through the subject. In a practical application, an X-ray detector is arranged to translate linearly on a track across the X-ray beam and at the end of the scan both X-ray tube and detector are rotated to a new angle and the linear motion repeated. Recent machines make use of a 'fan-beam' geometry, employing an array of detectors that detect simultaneously the X-rays arriving on a number of different paths through the subject.

Figure 9.4 shows the parallel geometry of the process. The subject (S) is shown as a fixed body having centred coordinates, x, y. The X-ray tube (X) and the single-point detector (D) rotate about the origin and translate (scan) across the subject. A second coordinate system, x', y', is also centred on S and is associated with the rotation angle θ of the detector position x' and the distance along the ray y'. Any point within the area S can be represented either by x, y or x', y', the two sets of

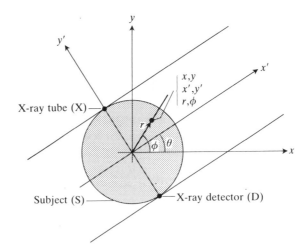

Fig. 9.4 Parallel geometry in tomography.

coordinates being related by a rotational transformation

$$x' = x \cos \theta + y \sin \theta$$
$$y' = -x \sin \theta + y \cos \theta. \tag{9.3}$$

If $u(x, y)$ represents the fraction of the X-rays absorbed by the subject (the *absorptivity factor*), with $u(x', y')$ as the corresponding function in the rotated coordinates, then the *projection* of the subject at this angle θ is defined as

$$p_\theta(x') = \int_L u_\theta(x', y') \, dy' \tag{9.4}$$

where L is the length of the beam path. This value $p_\theta(x')$ represents the data that are actually obtained at the detector at a given view angle θ. For each view angle $p_\theta(x')$ is sampled at a number of points along the x' axis, determined by the location of the detector (D) (actually a series of detectors would be used, having fixed locations along the x' axis). The angle θ is then incremented by a small amount and the process repeated. As many as 300 detectors and 360 view angles may be used, resulting in about $N = 10^5$ data points. This also corresponds roughly to the number of picture elements (*pixels*) displayed in a rectangular representation of the reconstructed image. If we consider the detected data as a vector X and the image data as another vector Y then the two are related by a matrix A having 10^{10} elements

$$X = A \cdot Y. \tag{9.5}$$

The problem lies in finding the inversion to this matrix A^{-1}, to obtain the image

$$Y = A^{-1} \cdot X. \tag{9.6}$$

Straightforward inversion is an enormous task requiring some N^3 operations. Clearly this is not practicable and tomography did not become feasible until a simpler alternative to matrix inversion was found. While iterative algebraic methods are used to reduce this amount of calculation [42], the most successful methods have been the use of Fourier transformation [43] (which was the first method to be used), and a variant of this known as *convolution-back-projection* [44]. The Fourier method will be considered first.

The 1-D Fourier transform of the projection $p_\theta(x')$ is given by

$$P_\theta(f) = \int_{-\infty}^{\infty} p_\theta(x') \exp(-j\pi x'f) \, dx' \tag{9.7}$$

where f is a frequency variable that can be positive or negative. If we now define a *two-dimensional* Fourier transform of $u(x, y)$, expressed in polar coordinates as $U(f, \theta)$, then an equality between this and $P_\theta(f)$ can be expressed as [43]

$$P_\theta(f) = U(f \cos \theta, f \sin \theta) = U'(f, \theta). \qquad (9.8)$$

This is known as the *projection theorem*, which states that the Fourier transform of a projection is the centre cross-section of the Fourier transform of the image. In principle, if $U(f, \theta)$ were known everywhere it would only be necessary to convert this to Cartesian coordinates $U_c(u, v)$ to reconstruct the slice of the subject (S) through a 2-D inverse transformation

$$u(x, y) = \int\!\!\!\int\limits_{-\infty}^{\infty} U_c(u, v)\exp[j2\pi(ux + uy)] \, du \, dv. \qquad (9.9)$$

Unfortunately $U(f, \theta)$ is not known everywhere but only at a finite discrete set of polar points (f, θ), so that in order to apply eqn (9.9) an interpolation to rectangular Cartesian coordinates needs to be formed. An exact interpolation method can be complex, and although methods exist for carrying this out directly in the frequency domain [45], most commercial machines work in the spatial domain through the convolution back projection approach.

In this technique the equivalent of eqn (9.9) is transformed to the spatial domain and the derivation of $u(x, y)$ obtained by convolution methods, which are relatively easy to approximate on a computer as 1-D finite sums. The two relevant equations are given below. The first is an *intermediate equation*

$$g_\theta(x) = \int_{-\infty}^{\infty} p_\theta(x')w(x - x') \, dx' \qquad (9.10)$$

which represents the result of a convolution of the projection of the image, at an angle θ and along a line x', with a window function $w(x - x')$. This is applied to a *back projection integral equation*

$$u(x, y) = \int_0^\pi g_\theta(r \cos(\theta - \phi)) \, d\theta \qquad (9.11)$$

which gives the absorptivity function required at x, y for a given angle θ. The operation represented by eqn (9.11) is called 'back projection', since it represents the integral of the actual view contribution of $g_\theta(r \cdot \cos(\theta - \phi))$ at each image point (x, y) or (r, ϕ). As can be seen from Fig. 9.4, the polar coordinates r, ϕ are related to x, y by $x = r \cdot \cos \phi$ and $y = r \cdot \sin \phi$. The window function $w(x - x')$ shown in eqn (9.10) is a simplified version of a complex function that carries out a process of interpolation for discrete data and a correction for the particular angle of view. A discussion of this and a fuller explanation of the back projection process

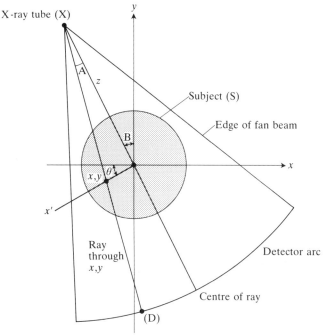

Fig. 9.5 Fan-beam geometry in tomography.

is given in [46]. An excellent bibliography on CT reconstruction methods is also included in this paper.

Most modern tomographic machines use a fan-beam geometry as shown in Fig. 9.5, which differs from the parallel-beam geometry by the need to modify the polar coordinates x' and θ into

$$x' = -z \sin A$$

and

$$\theta = B - A \tag{9.12}$$

where z is the distance between the X-ray source and the subject centre, with A and B relating to the detector angle and view angle of the fan beam, respectively. A conversion algorithm is described by Herman *et al.* [47], which obtains relationships very similar to that of eqns (9.10) and (9.11) except for an additional weighting factor included in the back-projection algorithm.

9.4.2 *Nuclear magnetic resonance*

Computer-assisted X-ray tomography has been an extremely successful technique for high quality representation of the interior of objects, particularly in the field of medicine. In the last decade other forms of tomography have been developed, either for special purposes or to overcome some of the practical difficulties of X-ray CT. One of the most successful of these has been the recent development of nuclear magnetic resonance (NMR), which gives a remarkable freedom from artifacts in the reconstituted image and, unlike X-ray CT, no ionizing radiation is used in the image formation. The NMR system is a purely electronic system relying for its operation on the interactions of rapidly varying magnetic fields with loosely bound hydrogen nuclei in the soft tissues of the body.

To understand NMR we have to appreciate the phenomenon of 'spin' in many atomic nuclei, particularly the hydrogen nuclei used for NMR. Such nuclei have an angular momentum arising from their inherent property of rotation or spin. Since nuclei are electrically charged this corresponds to a rotating electrical current, with which is associated a magnetic moment or dipole. In effect the hydrogen nuclei act as a multiplicity of tiny magnets, which can be affected by external magnetic forces. A constant magnetic force G will cause the dipoles to align themselves with the applied magnetic field. If a smaller alternating field B is now applied at right angles to this static field the magnetic dipoles will undergo a circular motion called *precession,* at an angle away from the normal aligned axis. It has been found that the frequency at which this precession is most pronounced corresponds to a particular resonant frequency f_L called the *Larmor* frequency, which is unique to one particular kind of nucleus. These frequencies are located in the *radio frequency* (RF) spectrum in the tens of MHz region. An indication of this resonant frequency is obtained by virtue of the small electromotive force $V(\omega)$ generated by the aligned and precessed group of magnetic dipoles, which may be detected directly the energizing RF field is switched off and before it decays to zero value.

This effect has been used to identify particular nuclei in a given chemical substance and provides biochemists with a useful tool for diagnostic spectroscopy. For NMR work however, only one nucleus is of major interest, namely hydrogen, which in the form of aqueous fluid makes up much of the body tissues. Some information on the general shape of a soft tissue organ where this differs from surrounding tissue can be obtained by Fourier transformation of the decaying RF signal $V(\omega)$, which carries out a conversion to the spatial domain, the x-axis across the subject area. But the interesting effect for CT purposes lies in the behaviour of the spinning dipoles following the removal of the RF field. The dipole precession gradually falls back to the axis of the static magnetic field and takes a finite time to do this, i.e. $V(\omega)$ decays slowly to zero amplitude value. It has been found that this delay time is related directly to the density of the body tissue. In a practical system three primary spatially varying physical properties are measured: the amplitude of the magnetic field, M_0 related to G and f_L, and two time delays, T_1 and T_2. These three values taken together enable a fairly detailed and multi-dimensional image of tissue structure to be obtained [32]. To reconstruct a 2-D image of the subject, the magnetic field gradient G is rotated to obtain a number of 1-D representations, one for each application of a pulse of RF energy at the nuclear resonant frequency. From a set of such data a cross-sectional image is reconstructed as described earlier.

9.4.3 Ultrasound

Another CT method is the use of *ultrasonic imaging* (USI) or *ultrasound*, which is also free from the use of ionizing radiation, or indeed any electromagnetic waveforms at all. The principle of use is very similar to that of sonar. It acts by propagating a disturbance of the body tissue, through which the acoustic ray travels. Various interactions with the tissue can be employed to differentiate between different layers. Examples are scattering, absorption, change in propagation speed, and Doppler frequency shift.

Although the earliest applications of USI made use of absorption of the acoustic waves in differing densities within the object, this was shown to give rise to considerable inaccuracies due to reflection, refraction, and diffraction. The most widely used method, which avoids most of these problems, is a spatial domain reconstruction from variations in velocity of the acoustic wave in the inhomogeneity [48]. The parameter measured is the time of flight of the ultrasound wave through the inhomogeneity. Figure 9.6 shows the geometry of the imaging system, which is essentially similar to Fig. 9.4. The acoustic waves travelling through the object are assumed to follow straight line paths p_j. For each angle of incidence, θ_i the receiver array measures the time of arrival of the waves for each of the rays J with respect to the time of transmission. This gives a set of time-of-flight measurements T_{ij}, which will differ depending on the velocity of the wave V_m in the medium surrounding the object (generally water) and the velocity V_0 within the acoustically transparent object. Denoting

$$n_k = 1/V_0 - 1/V_m \qquad (9.13)$$

where k is the cell number within the measured area, the problem is to find the variation in this parameter over a slice through the object, using many measurements of T_{ij} at various angles θ_i and ray paths p_j. A set of measurement equations is obtained

$$\sum_{k=1}^{N} F_{jk} n_k = z_{ij} \qquad (9.14)$$

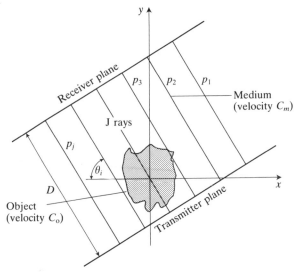

Fig. 9.6 Geometry of a system for ultrasonic tomography.

where F_{jk} is a function of the measured time of flight and the geometry of the ultrasonic system. This gives a set of J equations for each angle of incidence, θ_i and the problem reduces to the solution of a set of very large linear equations for these variables. The reconstruction problem is then identical to that discussed in connection with CT X-ray processing in Section 9.4.1.

9.4.4 Emission-computed tomography

Given suitable detection methods, reconstruction techniques may be applied to any 2-D representation of a subject that can be viewed from a multiplicity of angles. *Emission-computed tomography* (ECT) is an example where the derivation of a cross-section of the subject is obtained though internal radiation from the actual area to be scanned. Unlike the use of X-rays, which act as an external radiological source, the radiation detected in ECT derives from the immediate area of the subject being investigated. Suitable radiopharmaceuticals are ingested or otherwise located within the subject and their radiation is detected by means of an external scanning device. The method makes it easier to obtain multiple slices and allows real-time metabolic studies [49].

While convolution back-projection methods are fairly easy to apply, a major problem lies in tissue absorption for the radio emission in its path from the internal body source to the detector. Signal processing methods are applied to compensate for this, a typical algorithm being the transformation technique described by Bellini *et al.* [50]. The spatial resolution of ECT is lower than X-ray CT, since the amount of data collected is very much smaller. (Each radiation photon must be detected and analysed rather than the summation of the effects of many photons as in X-ray CT.) For this reason the reconstruction strategies incorporate iterative algorithms based on expected errors. Imaging time is also smaller due to isotope decay and movement of the radiochemicals within the subject, so that rapid on-line processing of detected information is needed.

In ECT we do not knw the exact position of the point source of radio emission, so it is necessary to define the direction of the ray proceeding from the source to the detector as well as its magnitude. In one form of ECT, *positron-emission-tomography* (PET), it is possible to take advantage of the mechanism of annihilation radiation that occurs close to the source. This releases two electro-magnetic photons in time coincidence, which proceed outwards in diametrically opposite directions. By placing two positron sensitive detectors on opposite sides of the source object the ray can be fully defined if the two detected events occur simultaneously. In practise a ring of such detectors is arranged around the object and these are scanned to detect radiation direction through the use of coincidence circuits. An alternative method, which is simpler but less sensitive, is to rely on gamma ray detection (which produces only one photon) and to use a collimation device to determine the direction [51]. In either case the detected information is applied to similar reconstruction algorithms to those described for X-ray CT.

References

1. Robb, R. A., Hoffman, E. A., Sinak, L. J., Harris, L. D., and Ritman, E. L. High speed X-ray computed tomography. *IEEE Proc.* **71** (3), 309–19 (1983).
2. Herman, G. T. and Lin, H. K. Display of three-dimensional information in computed tomography. *J. comput. assist. Tomog.* **1**, 155–60 (1977).
3. Bonner, R. E. Electrocardiogram monitoring by computer. In *Clinical electrocardiography and computers* (ed. C. A. Caceres and L. S. Dreifus). Academic Press, New York (1970).
4. Start, L., Okjima, M., and Whipple, C. H. Computer pattern recognition techniques. *Communs Ass. comput. Mach.* **5**, 10 (1962).
5. Rey, W. Adaptivity in cardiac monitoring: how to assess the measurement statistics, NTG Conference on Signalverarbeitung, Erlangen, pp. 160–8, April (1973).
6. Linkens, D. A. Empirical rules for the selection of parameters for autoregressive spectral analysis of biomedical rhythms. *Signal Process.* **1**, 243–58 (1979).
7. Pryor, T. A., Dazen, E., and Lakes, M., ed. *Computer systems for the*

processing of electrocardiograms. IEEE Computer Society Press, New York. Catalogue No. EH0170-1 (1980).

8. Pipberger, H. V., Boule, G., Berson, A. S., Briller, S. A., Geselowitz, D. B., Horan, L. G., and Schmitt, D. H. Recommendations for standardization of instruments in electrocardiography. Report of the subcommittee of the instrumentation committee on electrocardiography, American Heart Association. *IEEE Trans. biomed. Engng* **BME-14,** 60–8 (1967).

9. Ahmed, N., Milne, P. J., and Harris, S. G. Electrocardiographic data compression via orthogonal transorms. *IEEE Trans. biomed. Engng* **BME-22,** 484–7 (1975).

10. Schridhar, M. and Stevens, M. F. Analysis of ECG data for data compression. *Int. J. biomed. Comput.* **10,** 113–28 (1979).

11. Ahmed, N. and Rao, K. R. *Orthogonal transforms for digital signal processing.* Springer, Berlin (1975).

12. Hamba, S. and Techibara, Y. Representation and recognition of electrocardiograms using the finite Haar transform. *Electr. Engng Jpn* **96,** 111–7 (1976).

13. Kuklinski, W. S. Fast Walsh transform data compression algorithm: ECG applications. *Med. biol. Engng Comput.* **21,** 465–72 (1983).

14. Cain, M. F. and Ambosh, H. D. Fast Fourier transform analysis of signal averaged electrocardiogram. *Circulation* **69** (4), 711–20 (1984).

15. Beauchamp, K. G. *Applications of Walsh and related functions.* Academic Press, New York (1984).

16. Meffert, B., Schubert, D., Layarus, T., Poll, R., and Heusage, R. New and known methods of the application of transforms to quasiperiodic biomedical signals. IEEE Symposium of Electromagnetic Compatibility, Baltimore, pp. 336–41 (1980).

17. Rèmond, A., ed. *EGG informatics.* Elsevier, Amsterdam (1977).

18. Baldock, G. R. and Walter, W. G. A new electronic analyser. *Electronic Eng.* **18,** 339–44 (1946).

19. Dumermuth, G. and Fluhler, H. Some modern aspects in numerical spectral analysis of multi-channel electroencephalographic data. *Med. biol. Engng* **5,** 319–31 (1967).

20. Gevins, A. S. and Yeager, C. L. EEG spectral analysis in real times. *DECUS Proc.* **Spring,** 71–80 (1972).

21. Smith, W. D. Walsh versus Fourier estimators of the EEG power spectrum. *IEEE Trans. biomed. Eng* **BMW-28,** 790–3 (1981).

22. Welch, P. D. The use of fast Fourier transform for the estimation of power spectra; a method based on short, modified periodograms. *IEEE Trans. Audio Electroacoust.* **AU-15,** 70–3 (1967).

23. Weide, B., Andrews, L. T., and Iaunone, A. M. Real-time analysis of EEG using Walsh transforms. *Comput. Biol. Med.* **8,** 255–63 (1978).

24. Setton, J. J. and Smith, W. D. An EEG monitor using the Walsh transform on a standard microprocessor. *IEEE Trans. biomed. Engng* **BME-26,** 525 (1979).

25. Larsen, H. and Lai, D. C. Walsh spectral estimates with applications to the classification of EEG signals. *IEEE Trans. biomed. Engng* **BME-27,** 485–92 (1980).

26. Jansen, B. H. Walsh spectral estimates with application to the classification of EEG. *IEEE Trans. biomed. Engng* **BME-28** (12), 836–8 (1981).

27. Ahlbom, G. and Zetterberg, L. H. A comparative study of five methods for the analysis of EEG. Ray Institute of Technology, Technical Report No. 112, Stockholm (1976).

28. Dumermuth, G., Gasser, T., and Lange, B. Aspects of EEG analysis in the frequency domain. In *Computerized EEG analysis* (ed. G. Dolce and H. Kunkel) pp. 432–57. Gustave Fisher, Stuttgart (1975).

29. Bates, R. H. T., Garden, K. L., and Peters, T. M. Overview of computerized tomography with emphasis on future developments. *IEEE Proc.* **71**, 356–72 (1983).

30. Cho, Z. H. and Burger, J. R. Construction restoration and enhancement of 2- and 3-dimensional images. *IEEE Trans. nucl. Sc.* **NS-24** (2), 886–99 (1977).

31. Budinger, T. F., Gullberg, G. T., and Huesman, R. H. Emission computed tomography. In *Image reconstructions from projections* (ed. G. T. Herman) pp. 147–246. Springer, New York (1979).

32. Hinshaw, W. S. and Lent, A. H. An introduction to NMR imaging: from the Bloch equation to the imaging equation. *IEEE Proc.* **71**, 338–50 (1983).

33. Greenleaf, J. F. Computerized tomography with ultrasound. *IEEE Proc.* **71**, 330–77 (1983).

34. Raviv, T., Greenleaf, J. F., and Herman, G. T., ed. *Computer-aided tomography and ultrasonics in medicine.* North Holland, Amsterdam (1970).

35. Bates, R. H. T. Full-wave computed tomography. Part 1—fundamental theory. *IEE Proc.* **131A** (8), 610–15 (1984).

36. Seager, A. D., Yeo, T. S., and Bates, R. H. T. Full-wave computed tomography. Part 2—resolution limits. *IEE Proc.* **31A** (8), 616–22 (1984).

37. Herman, G. T., ed. *Image reconstruction from projections.* Springer, Berlin (1979).

38. Kniss, J. P. Radioisotope scanning in medical diagnosis. *Ann. Rev. Med.* **1**, 381 (1963).

39. Picer, S. M. and Vetter, H. G. Processing of radioisotope scans. *Nucl. Med.* **10**, 150–4 (1968).

40. King, R. E. Digital image processing in radioisotope scanning. *IEEE Trans. biomed. Engng* **BME-21**, 414–16 (1974).

41. Hunt, B. R., Jauney, D. H., and Zeigler, R. K. Digital processing of scintigraphic images. IBM Wissenschaftliches Zentrum Technical Report No. 70-03-001, Heidelberg, March (1970).

42. Gordon, R., Bender, R., and Herman, G. T. Algebraic reconstruction techniques (ART) for three-dimensional electron microscopy and X-ray tomography. *J. theor. Biol* **29**, 471–81 (1970).

43. Bracewell, R. N. Strip integration in radioastronomy. *Aust. J. Phys.* **9**, 198–217 (1956).

44. Shepp, L. A. and Logan, B. F. The Fourier reconstruction of a head section. *IEEE Trans. nucl. Sc.* **NS-21**, 21–42 (1974).

45. Stark, H., Woods, J. W., Paul, I., and Hingorani, R. Direct Fourier reconstruction in computer tomography. *IEEE Accoust. Speech Signal Process.* **ASSP-29** (2), 237–45 (1981).

46. Lewitt, R. M. Reconstruction algorithms: transform methods. *IEEE Proc.* **71** (3), 390–408 (1977).

47. Herman, G. J., Lakshminarayan, A. V. and Naparstek, A. Reconstruction using divergent ray shadowgraphs, *Comput. Biol. Med.* **6**, 259 (1976).

48. Greenleaf, J. F., Ritman, E. L., Wood, E. H., Robb, R. A., and Johnson, S. A. Algebraic reconstruction of spatial distributions of acoustic velocities in tissue from their time of flight profiles. In *Acoustical holography* (ed. P. S. Green) Vol. 6, pp. 71–90. Plenum Press, New York (1975).
49. Budinger, T. F., Gullberg, G. T. and Huesman, R. H. Emission-computed tomography. In *Image reconstruction from projection* (ed. G. T. Hermann) pp. 147–246. Springer, New York (1979).
50. Bellini, S., Cafforio, C., Piacentini, Mard and Rocca, F. Design of a computerized emission tomographic system. In *Digital signal processing* (ed. V. Cappellini and A. G. Constantinides) pp. 207–16. Academic Press, New York (1980).
54. Knoll, G. F. Single photon emission computed tomography. *IEEE Proc.* **71** (3), 320–9 (1983).

10

IMAGE PROCESSING

10.1 Introduction

A major application of domain transformation techniques lies in the manipulation of multi-dimensional signals. Large amounts of data are generally involved, so that computational economy and speed in carrying out the individual operations are paramount to successful processing.

Several examples of 2-D processing have already been given in earlier chapters and the purpose here is to review broadly the general principles involved and to note the use made of transformations in the range of techniques currently being applied. The discussion will be limited to 2-D or image processing. Examples are found in facsimile transmission, television images, medical X-ray photographs, radar and sonar maps, satellite transmitted pictures, electron microscope records, seismic data displays, and in many other applications. For most purposes a picture is considered as divided into a matrix of sub-pictures, known as *picture elements* or *pixels,* in which the average brightness intensity is quantized and expressed as a digital number. This sampled and digitized form of a picture is referred to as an *image* or image matrix.

There are many reasons for wanting to process this image. The clarity of the image may have suffered during its acquisition or in the process of transmission. One example of this is the enhancement of X-ray images through tomography, where the view we wish to observe is obscured by other structures. Another is the improvement of satellite transmitted pictures to enable small details to be observed despite the noise imposed by the mechanism of data transmission. We may wish to select particular aspects of a picture for emphasis or detailed examination, such as the outlining of a set of chromosomes as seen through a microscope or a partially hidden military tank viewed from a satellite. We call the reduction of image degradation *image restoration* and the improvement of the subjective aspects of a picture *image enhancement.*

An extension of this kind of selection would be the classification of objects contained in a given picture. This leads to the idea of *pattern recognition,* i.e. the detection and extraction of particular patterns or features from an image for the purpose of classification or for facilitating easier recognition of some identifiable feature. This covers a wide range of applications including automatic written character recognition, iden-

tification of man-made structures in a landscape, and equipping robots
with vision capability.

The sheer volume of digital data images creates difficulties in storing
high definition images or transmitting these over a limited bandwidth
communication link. For a reasonable definition we will be considering
up to 10^6 sampled values for each constituent image. Thus there is an
interest in *data compression* methods, i.e. ways of compressing the data
into a smaller volume or occupying a smaller transmission bandwidth
with a minimum penalty in the reduction of definition for the reconstructed image.

In all these application areas domain transformation has an important
role to play and a brief look at how this is applied in a number of image
processing situations will now be given. Other techniques based on
optical methods, statistical operations, digital filtering or the various
forms of pulse encoding for transmission fall outside the scope of this
book and are fully discussed elsewhere [1, 2, 3], as are holography and
the subject of computer graphics, which deals with manipulation of
images created specifically for computer display.

10.1.1 Some basic definitions

Digitization of a picture presents similar problems in the choice of
sampling methods and quantization levels as in the 1-D case. There are
however a number of additive factors resulting from the subjective nature
of most reconstructed images. A method of sampling and quantization
suitable for one particular image and application may be subjectively
inadequate in another context.

Subdivision of a picture into pixels is usually taken at a regular matrix
spacing, but there may be advantages in non-uniform spacing (to correct,
for example, perspective errors introduced by the acquisition device, e.g.
camera tilt). Representation of the pixels can be taken as the average
grey level at the actual point of sampling, although other forms of level
representation have been applied. For many processing purposes it is
useful to divide the image matrix into sub-matrices containing n^2 pixels in
order to improve coding efficiency [4]. The choice of sub-matrix size is
dependent on the type of transformation selected, and will be found to lie
between $n = 4$ and $n = 16$. While the MSE should improve with increasing n as the number of correlations taken into account also increases, it
has been found that most pictures contain significant correlation between
adjacent pixels, and no substantial improvement in performance is found
for $n > 8$ [4, 5].

Sampling of a picture means applying a sampling function to select
values of the picture intensity level at discrete points. At all other points
the sampled image value will be set to zero. The sampling is carried out

at regularly spaced intervals along the x axis using a function

$$s(x) = \sum_{n=-\infty}^{\infty} \delta(x - n) \qquad (10.1)$$

where $\delta(x - n)$ is a Delta impulse function with unity spacing for the impulse sample.

In practice the sampling function $s(x, y)$ consists of an array of impulses spaced δx in the x direction and δy in the y direction and we can write

$$f_s(x, y) = \sum_{i=-\infty}^{\infty} \sum f(x_i, y_i)\, \delta(x_i - i\,\delta x, y_i - i\,\delta y). \qquad (10.2)$$

Synthesis of a sampled image is generally obtained by linear interpolation of the grey levels between sampled points but other forms of interpolation are possible, such as sinusoidal interpolation. As with 1-D sampling, the sampling theorem is applicable. For a 2-D image $f(x, y)$ having the maximum spatial frequencies of f_1 for x and f_2 for y, sampling frequencies of $f_{s1} \geqslant 2f_1$ and $f_{s2} \geqslant 2f_2$ are required. Usually a unique value of f_s is chosen equal to the greater of f_{s1} and f_{s2}, with a corresponding sampling interval of $h = 1/f_s$.

Of course, sampling the image will give rise to similar aliasing situations to those found with 1-D sampling, and it will be necessary to limit the bandwidth before sampling in much the same way through the use of a rectangular window in the frequency domain (see Section 1.2). Aliasing of images results in Moire patterns in the reconstructed image, which can only be avoided by increasing the sampling rate.

Subject to these constraints it should be possible to fully recover the original image through convolution of the sampled image $f_s(x, y)$ with the spatial form of this rectangular window. However to do this completely an unlimited area of sampled image needs to be available. Since this is not possible some distortion in the reconstruction process occurs, requiring further modification, equivalent to windowing in the 1-D case (Section 1.3.2). Quantization of the sampled image data refers to the assignment of a particular grey level for a data sample. As in the 1-D case, if q is the quantization step then the image grey level range is represented through Nq discrete grey levels, where N is a suitable integer (often 256). The levels are generally equal but can follow a logarithmic or some other function base. A pseudo-random quantization has been proposed to remove an undesirable outline of the picture content. A variable rate dependant on the picture content is also used [6].

To put all of this into formal terms we represent the original image as a real-valued function of two variables $f(x, y)$ expressed on a plane domain in terms of Cartesian coordinates, and define the digitized image as a series of real numbers $x_{i,k}$. These can be expressed in matrix form as $A_{i,k}$, where $i, k = 1, 2, \ldots, N$. Digitization is the construction of a sequence of real numbers $A_{i,k}$ from image information $f(x, y)$ such that the image information is preserved within a specified error criterion. Mathematically the process can be defined as

$$A_{i,k} = F[f(x, y)[P]]\qquad(10.3)$$

where $[P]$ is a finite parameter set and F is a transformation procedure.

Similarly, the reconstruction of a picture from the digitized sequence $A_{i,k}$ can be obtained through a further transformation G to obtain a 2-D image, $g(x, y)$

$$g(x, y) = G([A_{i,k}]x, y).\qquad(10.4)$$

The error may be defined in terms of a comparison method as

$$f(x, y) \copyright g(x, y) \leqslant e(x, y)\qquad(10.5)$$

If the comparison method $\copyright$ is a MSE one, then

$$\int\!\!\int_{-\infty}^{\infty} [f(x, y) - g(x, y)]^2 \, dx \, dy \leqslant e(x, y)\qquad(10.6)$$

The object of image reconstruction is to reduce $e(x, y)$ to a minimum. To do this a process of image restoration is carried out, as described later. First however it is necessary to consider the general behaviour of a system carrying out this process of converting a picture into a digitized image. This is shown in Fig. 10.1.

The performance of a 2-D system is described in similar terms to a 1-D system. The frequency response function is replaced by a *modulation transfer function,* which gives the modulus of the Fourier transform of a 2-D matrix present at the output of a system, thus giving a measure to the attenuation of the sinusoidal variation in the image content. Similarly the *phase transfer function* is the phase of the Fourier transform of the output matrix and corresponds to the 1-D phase response of a system.

By far the most important of these concepts is that of the *point spread function* (PSF). This is equivalent to the impulse response of a 2-D system and would describe for example the distortion imposed by an optical system interposed between the object and its image. If we let $o(u, v)$ be a 2-D function describing the object of interest, then its image

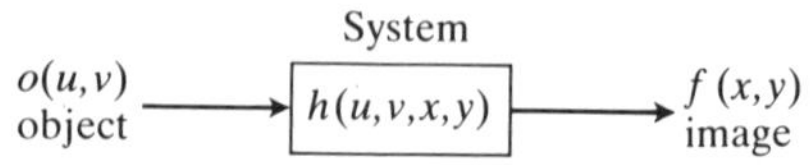

Fig. 10.1 The point spread function.

$f(x, y)$ can be stated as

$$f(x, y) = \int\!\!\int_{\infty}^{\infty} h(u, v, x, y) o(u, v) \, \mathrm{d}u \, \mathrm{d}v \qquad (10.7)$$

where (u, v) and (x, y) represent the spatial coordinates of the object and its image, respectively.

$h(u, v, x, y)$ is the PSF of the system producing the image, $f(x, y)$. It can be shown that the image we obtain is really the convolution of the object with the PSF describing the system. This has certain implications in image reconstruction that will be referred to later. Such systems are called *linear isoplanar systems*.

2-D transformations were considered in Sections 2.5 and 5.6. It will be assumed in the following applications that these are carried out by a sequence of 1-D transformations, first of the rows and then of the columns of the image matrix [4, 7]. Orthogonal and unitary transformations are also assumed where the choice is limited to those having a fast transformation algorithm. This excludes the KLT and the *singular value decomposition* of the image (SVD), both of which require considerable computation capability of at least N^3 complex products [8].

10.2 Image compression and transmission

Image compression is required for the transmission of television, facsimile or videophone information, where economy in transmission bandwidth is important. It is also needed to reduce the volume of stored image data such as X-ray images and satellite transmitted pictures. The principles of data compression for 1-D information were considered in Section 7.4.1. Similar principles are involved in image compression.

Consider first the transmission of a television image. Two general approaches for image compression have been investigated. One uses spatial domain techniques, such as *intra-frame coding*, utilizing the considerable redundancy in adjacent picture frames [9], and the other uses transformation techniques. This second approach is in more general use and will be considered first.

The two factors looked for in image coding are compressional efficiency and ease of computation. At present the cosine transform provides the best approach to these two factors.

As discussed in Section 5.6.1, fast transformations for the cosine transform were initially implemented by using the FFT as a basic algorithm [10]. This method employs a double size FFT of $2N$ coefficients and requires complex arithmetic. An alternative algorithm, known as a the fast discrete cosine transform (FDCT), requiring only real operations, is also available and is approximately six times as fast as the FFT implementation [11]. This method can be used with fast sequency transforms and several Walsh transform implementations of the fast cosine transform have been used [12, 13].

Somewhat simpler hardware construction is required for a particular approximation to the cosine transform, known as the C matrix transform and described by Srinivassan and Rao and [14] which has the advantage that the conversion matrix has only integers as its elements.

A sine transform having similar capabilities for image processing has also been defined, but it has less efficient compressional characteristics [15]. It can however yield a faster computation algorithm than the DCT. A useful review of its performance and implementation is given by Yip and Rao [16]. A method of deriving this transform from the cosine transform is described by Wang [17].

The value of the cosine transformation for image compression lies in its good variation distribution and low rate of distortion function. This results in efficient energy compaction, where the transform coefficients containing the largest variances are found to be contained in approximately a quarter of the transformed matrix. Its variance characteristics also lend themselves to efficient *adaptive coding compression* techniques, which are based on the statistics of the cosine transformed image. In a scheme described by Chen and Smith [18], transformed sub-pictures are sorted into classes by the level of image activity present. Within each activity coding, bits are allocated to individual transform elements according to the variance matrix of the transformed data. Bits are then distributed between the high-activity (busy) and low-activity (quiet) areas, with few bits allocated to the quiet areas. A bit rate of a 1 bit per pixel for a monochrome image is obtained in this application. A similar adaptive bit coding has also been applied to sub-matrix coding, where the optimum picture compression is found to differ from that applied to the complete image matrix [19].

10.2.1 Image transmission

The transmission of still and moving images is a major application for image compression. Until fairly recently the transmission of still images has meant facsimile transmission in which image reconstruction can take several minutes and no advantage is realized in using compressional techniques. However, with the growing use of videotext, the transmission of still images is now required in seconds. Because of the slow transmission rates presently in use, it is essential that some form of image compression be used.

The transmission of moving images (e.g. television) is much faster, and although a wider bandwidth is allowed, there are still considerable advantages in using data compression, particuarly for satellite-transmitted pictures. Actually, in terms of the product of bandwidth and reconstruction time, the two cases are very similar. Presently, videotext uses a transmission rate of 4 800 bit s^{-1} or less with about 6 s for reconstruction, giving a product of approximately 30 000 bit s^{-1}. Television transmission of 626-line standard requires a frame reconstruction time of 1/50 s and, assuming 8-bit coding to provide full luminance and colour gradation, the product is approximately 60 000 bit s^{-1}. Hence we find that similar image compression methods are used in both applications.

The transmission of the transform of an image rather than the image itself can lead to considerable opportunities for reduction in transmission bandwidth due to the effective energy compaction of orthogonal transformation [20]. All unitary transforms are information-preserving however, so that no bandwidth reduction results from simply transmitting the transform of an image. Instead the non-zero transform coefficients are restricted in number before transmission takes place, so that upon reconstruction of the now limited bandwidth image a degraded version of the original image can be realized. Figure 10.2 illustrates the general principle. The image $f(x.y)$ is transformed through a unitary transformation to yield $F(x, y)$. The transformed matrix $A_{i,k}$ is then subjected to some form of selective sampling function $S(x, y)$, which carries out the data compression to give a sampled and compressed signal

$$F_s(u, v) = F(x, y) . S(x, y). \tag{10.8}$$

The sampling function takes on a value of 0 or 1 according to some prior rule and so reduces the number of finite transformed values $F_s(u, v)$ for transmission. This can be simple threshold selection, in which coefficients smaller than a given value are made zero, or more complex methods can be used. Also of interest in transform compression is the question of block size. For a 1-D signal this is simply the number of elements in the transformed signal and the performance of the transformation increases with the number of elements transformed. With 2-D transmission, while the performance increases with the number of elements transformed, intermediate memory storage is used to store the transformed coefficients in the first direction while the transform is being computed in the second

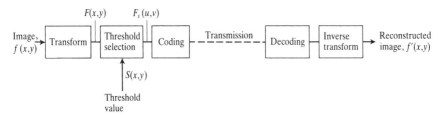

Fig. 10.2 Transform image compression.

direction. Hence it is usual to partition the image into a number of sub-images and to transmit the complete image on a block-by-block basis to optimize the memory required [21].

For transmission across a given communication channel it is necessary to code the compressed digital data. The various methods of coding used to minimize error in the transmission process will not be considered here. Following the decoding procedure the recovered data has zeros added at pixel locations corresponding to those below a threshold value, and the augmented signal is then inversely transformed to give an approximation to the original image, $f'(x, y)$.

10.2.2   Television data compression

Data compression techniques for moving television images are successful because of the high levels of correlation that occur between the grey levels of spatially adjacent picture elements. This provides a redundancy that can be exploited in intra-frame coding, with transform coding forming one important technique [22]. The reduction in transmitted coefficients is obtained by the threshold elimination mentioned earlier, in which all those coefficients that fall below a specified value are discarded. Other methods of selection are used and described in the literature [11, 20].

Another source of redundancy that can be exploited relates to the high correlation that exists between subsequent frames of the transmitted television picture. This is known as *inter-frame coding*.

A data compression technique for intra-frame colour television transmission has been described by Jalali and Rao [23]. This sub-divides the picture into a number of sub-images of 4, 8, or 16 pixels, which are transformed separately through the FDCT and the significant coefficients only transmitted. By adopting a highly modular structure in a pipeline configuration it has been possible to maintain a throughput rate of the NTSC television carrier at three times the sub-carrier frequency of 10.7 MHz.

Most inter-frame compression systems use 2-D transform coding, often utilizing a cosine transform in conjunction with *differential pulse code modulation* (DPCM) between successive frames. This technique encodes the difference between a given transform coefficient and the corresponding encoded coefficient from the preceding scan line for the frames of an interlaced scanning system. There is usually built in to the DPCM feedback loop an attenuation factor that diminishes the value accumulated from previous entries prior to their combination with a new value. This practice provides some immunity against transmission error by including a self-correcting feature for synchronization faults. A survey of such systems is given by Kamangar and Rao [24], in which the use of Hadamard, cosine, and Fourier transformations are compared.

10.3 Image restoration and enhancement

In *image restoration* the object is to compensate for the degradations introduced into a picture by the acquisition, transmission, and imaging process. Typical corrections to the image include removal of random noise, often inherent in the means of image acquisition, and correction for various optical or other aberrations such as motion blur and defocussing.

In *image enhancement* the image is modified either to assist human viewing or to facilitate further machine processing. Computer enhancement of a digital image is generally a subjective process, since the object is to emphasize or reveal some aspect of the image not immediately apparent by visual inspection. Three important categories considered are intensity mapping, edge detection, and pseudo-colour. Tomography, discussed in the previous chapter, may be considered as a more complex example of image enhancement using a series of multiple images.

There are no generalized approaches based on a theory of enhancement or restoration. Instead a number of techniques have been developed that are useful in a number of processing situations. Some of these are considered below.

10.3.1 Image restoration

Restoration is the reconstruction of the image by inversion of some degradation phenomena, and for this to be successful some knowledge concerning the degradation characteristics will be required. This may take the form of analytic or statistical models or other *a priori* information based on the known information about the image structure. Since optical signal images have the special property of always being positive quantities they do not obey completely normal statistical operations, so image restoration can be considered as an *ill-conditioned process* and as such lacks a unique solution. For this reason we find a number of empirical methods in use, although some broad trends in image treatment are discernible. The restoration process often is not a linear one, and it is common to find that non-linear methods produce superior results to linear ones [25]. Most degradations will however be found to be linear superpositions of the desired image with distortions and a noise factor. This implies distortion in the spatial frequency spectrum, so that one way of restoring the original image is through modification of the spectrum of the degraded image. This can carried out by spatial filtering or by convolving the image with a PSF in an image construction process.

For linear shift-invariant phenomena such as distortion caused by atmospheric turbulence or optical camera blur, the image can be restored through the use of *inverse filtering*. For this we require a knowledge of the PSF of the system and its transform, referred to as the *frequency transfer function*, $H(\omega)$. An inverse filter corresponding to this distorted system may be defined as a filter having a frequency response of

$$F(\omega) = 1/H(\omega) = H^{-1}(\omega). \tag{10.9}$$

The situation is illustrated in Fig. 10.3, from which it is clear that the output $o(x, y)$ of the inverse filter will be theoretically equivalent to $i(x, y)$, the true input image in the absence of noise. In a practical case however, additive noise, $n(x, y)$ will be introduced and the reconstructed image in the frequency domain will be,

$$O(\omega) = [I(\omega)/H(\omega)) - (N(\omega)/H(\omega)] \quad \text{for } H(\omega) \neq 0. \tag{10.10}$$

If the noise is large, separate filtering is required to reduce the effect of $n(x, y)$ before inverse filtering takes place. Where the SNR is high over the image bandwidth, inverse filtering can yield quite good results [26].

Several problems are present with this simple model for inverse filtering. It may be difficult to obtain a value for the PSF if *a priori* information on the system characteristics is less than complete. Also, an inverse filter may not exist due to zeros in $H(\omega)$, with subsequent instability if its realization is attempted.

Wiener filter restoration requires extensive *a priori* information on the PSF in order that its transform may be computed for the filter [27]. Statistical methods such as Bayes estimation for image enhancement also require *a priori* statistics of the picture and noise [28, 29].

Simple methods of image restoration that do not require knowledge of the PSF may be based on threshold filtering. This is applicable for example in the case of smoothing or reduction of the effect of ragged edges or noise in an image. Frequency domain noise tends to occupy the higher frequency regions so that simply smoothing can in theory improve

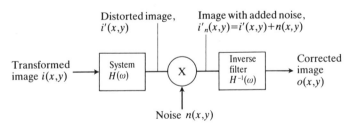

Fig. 10.3 Inverse filtering.

the subjective appearance of an image. This may be achieved through low-pass filtering in the frequency domain, i.e. applying a threshold value to the transformed image coefficients, reducing those above a given frequency value to zero. A difficulty with this procedure is that it is equivalent to convolution of the spatial image with a $(\sin x)/x$ function and so introduces Moire patterns or 'ringing' in the inverse transformed image. Whilst this can be avoided by careful shaping of the threshold transfer function, such as the use of a Gaussian transfer function (remembering that the Fourier transform of a Gaussian function is itself a Gaussian function and hence does not alter its shape in the frequency domain), this is likely to reduce the sharpness of detail in the image as well as reducing the noise.

A way to avoid the arbitrariness of the inverse filtering approach is to apply a MSE criterion to the derivation of the inverse filter function [30]. The minimum MSE criterion is one of the few objective criteria for which the optimum restoration can be computed. By making certain assumptions that the noise and signal are both stationary random processes and exhibit a Gaussian characteristic over the entire 2-D plane, then it can be shown that an optimum MSE performance is obtained by using an inverse filter function

$$F'(\omega) = \frac{H\star(\omega)}{|H(\omega)|^2 + P_{N(\omega)}/P_{I(\omega)}} \tag{10.11}$$

where $P_{I(\omega)}$ and $P_{N(\omega)}$ are the power spectral density functions of the image and noise signals, respectively. For the most part $P_{N(\omega)}/P_{I(\omega)}$ is assumed constant over the frequency range of interest, although for some applications Horner has suggested that a negative exponential shape for this function may be more appropriate [31]. Note that with $P_{N(\omega)} = 0$ then $F'(\omega)$ becomes an inverse filter $F(\omega)$, and that when $H(\omega) = 0$ then $F'(\omega) = 0$ instead of infinity as in the case of the inverse filter.

For certain degradation characteristics, such as optical camera blur, a deconvolution technique based on the cepstrum has proved effective. A blurred image can be regarded as a convolved set of identical images, each displaced slightly along the x-axis. Since the convolution effects of the PSF are additive in the cepstral (pseudo-spatial) domain then separation, and hence spatial filtering of the effects of blur, becomes possible [32].

Transform techniques dependent on adaptive 2-D filters and for which prior knowledge of the PSF is not required are also described by Smith [33]. This method carries out Walsh sequency image filtering using a procedure designed to eliminate statistically insignificant Walsh power spectra before inverse transformation of the data to restore the modified image. Other methods not dependent on *a priori* information include maximum entropy and homomorphic techniques, but a discussion of these and the ways of deriving an acceptable PSF for a given image fall outside the scope of this book (see [34, 35]).

10.3.2 Intensity mapping

Intensity mapping is generally a non-linear operation

$$f'(x, y) = I(f(x, y)) \qquad (10.12)$$

where $I(.)$ is a non-linear mapping of $f(.)$ independent of the intensity values of $f(x, y)$ in the picture content. Often the grey levels are not well distributed by the quantizing process and two methods of non-linear mapping to correct this are found useful. The first is to use a logarithmic device to stretch the quantization over the grey scale to enhance a particular grey region [36]. The second is to use a histogram equalization scheme in order to produce a result as close as possible to a uniform intensity distribution. In effect as many shades of grey as possible are used to display the image. Equalization of this type is particularly valuable with X-ray images and other images showing a histogram heavily biascd towards one end of the grey scale [37].

10.3.3 Edge detection

Edge detection is a useful first step in extracting information from a recorded image and an essential preliminary for shape determination, considered later in this chapter. Several methods have been studied for automatic digital edge extraction. Some of these are:

(a) two-dimensional frequency or sequency filtering;
(b) local operations through gradient detection;
(c) contrast enhancement by transformation;
(d) statistical and relaxation methods.

An edge to an image corresponds to high frequency or high sequency components in the x and y directions. A high-pass or band-pass digital filter may be applied to the image and a contrast expansion obtained. This method does not perform well on noisy images however, since the noise itself consists mainly of higher frequency components. When this method is used there are advantagess in the use of a Walsh or Haar function series in the sequency filtering, since these functions match more closely the characteristics of the edge discontinuity [38, 39].

An edge is a region of the image where the intensity. gradient has a large magnitude. Various gradient operators have been devised to detect this [1], and in some cases these have been combined with directional information to permit edge following to take place in the manner of *contour tracking* [40]. The idea of tracing contours of a constant grey level is useful as a means of reducing the quantity of data that needs to be handled or stored. If a contour is followed, tracing the constant grey level values, then only the coordinate addressing data together with grey level value is required to be retained from the processing operation.

A well-known local operator for edge detection is the *Sobel operator* [41]. Consider a 3×3 block of pixels given by

$$
\begin{array}{ccc}
a_0 & a_1 & a_2 \\
a_7 & f(x, y) & a_3 \\
a_6 & a_5 & a_4.
\end{array}
\tag{10.13}
$$

This consists of the current central pixel, $f(x, y)$, surrounded by eight neighbouring pixels numbered 0 to 7. A new averaged value is calculated for $f(x, y)$ in terms of the neighbouring pixels as,

$$
f'(x, y) = \sqrt{(X^2 + Y^2)} \tag{10.14}
$$

where

$$
\begin{aligned}
X &= [a_2 + 2a_3 + a_4] - [a_0 + 2a_7 + a_6] \\
Y &= [a_0 + 2a_1 + a_2] - [a_6 + 2a_5 + a_4].
\end{aligned}
\tag{10.15}
$$

Equation 10.14 is repeated for each new value of $f(x, y)$ over the entire image, usually in conjunction with a threshold operation on $f'(x, y)$, to enhance the changes in gradients contained within the picture. Some increase in noise level of the processed image is inevitable with this method.

Using transformation methods the sequency functions are significant owing to the match of these functions to the abrupt discontinuities looked for. This is particularly the case with the Haar function and its double-step characteristic. Generally orthogonal transforms are global in concept since they involve processing all the pixels contained in a sub-image. There is no inherent selection of the higher spatial frequency components, which contain edge representation. The Haar transform, on the other hand, has the majority of its basis functions representing well-defined edges in the spatial domain as edges in the transform domain, i.e. it is a local rather than a global transform. Because of this local processing characteristic, the Haar transform is ideal to use in conjunction with thresholding or in the selection of the n largest coefficients. This process has the effect of an adaptive compression of the coefficients in which the bit compression is relatively low in areas of high image activity (at the picture contours for example) and high in other areas. Examples of the improvement in image compression for the Haar transform in this application are given by Lynch and Reis [42], who show that for edge representation in a 4×4 image array the Haar transform requires at most 10 coefficients for negligible error, whereas the global

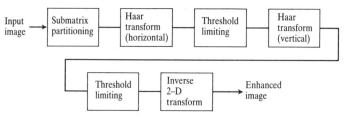

Fig. 10.4 Threshold edge enhancement.

transforms in a similar situation require all 16 coefficients. This has also been noted by Sivak [43, 44], who applies threshold selection to infra-red images. In this application the four highest spatial sequency coefficients from each sub-matrix are retained and others reduced to small values by means of a weighting algorithm. This is illustrated in Fig. 10.4. First the horizontal lines of the digital image are transformed through a 1-D FHT and then subject to a threshold operation to enhance vertical edges. Similar operations on the vertical lines are carried out to enhance horizontal edges. Finally inverse 2-D transformation recovers the spatial image having enhanced scene edges.

Statistical and relaxation methods are reviewed by Levialli [45]. Several of these use the concept of *masks,* in which the image is divided into sub-images containing a small number of pixels. These are multiplied with the corresponding pixels in a defining mask or series of mask images. These contain weighting coefficients appropriate to an edge gradient having a particular direction through the sub-image (e.g. horizontal, vertical, diagonal, and having positive or negative values to simulate edge polarity). The summation of these pixel value products (sub-image and mask) gives an indication of the probability of the defined edge being present [46]. Often this indication is combined with other local tests such as threshold value and direction, after which a binary decision is taken, which may be incorporated into a contour edge map (Fig. 10.5). Mask edge detection is also proposed by Shore in which selected 2-D Haar functions are used to give an edge direction indication [47].

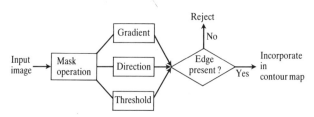

Fig. 10.5 Multi-decision edge detection.

10.3.4 Pseudo-colour enhancement

The objective of *pseudo-colour enhancement* is to increase the effective viewing dynamic range of the original grey scale by appealing to the human visual response to colour. This is achieved by making a colour image from a monochrome image, mapping for example a particular spatial frequency range to a particular colour shade, or mapping a particular grey shade to a given intensity hue defining a colour shade. The amount of information that can be conveyed to the observer in this way can be quite considerable and the method has the ability to emphasize aspects of the original picture that are by no means apparent from the study of its detail.

A number of methods of implementation for these concepts are possible. One very simple technique that does not mix the primary colours involves mapping a separate portion of the grey scale into maximum brightness for each primary component. Examples of this technique are given by Andrews [48] and demonstrate the difference between the X-ray images of a normal and a diseased lung when pseudo-colour is used. A somewhat more complex pseudo-colour generation technique involves mixing of the primary colour components in accordance with a linear mapping. Here the pure primary colours occur only at the extreme ends and at the exact centre of the grey scale. Another technique is to use a variable position window that colours all the grey levels within that window one colour, leaving the rest of the image untouched. This would be useful for example in coloured contour mapping of the kind we find in conventional geographic maps.

10.4 Pattern recognition

Three major approaches to pattern recognition can be distinguished. They are:

1. *Template matching,* where the shape of the class of pattern is known [49].

2. *Decision theoretic* or statistical method, in which we look for characteristic measurements, called features, in the image and attempt to classify the image from these measurements [50, 51].

3. *Syntactic* or linguistic approach, in which a hierarchical information structure is produced. The pattern may be classified in terms of simpler sub-patterns and each simpler sub-pattern can again be described in terms of even simpler sub-patterns, and so on [51–53].

10.4.1 Template matching

In the template matching process a set of templates, one for each pattern class, is stored in the computer memory. The signal image is then matched or compared with each of these templates and a classification made based on a given matching criteria (e.g. value of a correlation coefficient). The templates may be stored directly as 2-D images for each class requiring identification, or alternatively the transformation of the template image may be stored. The latter often results in a smaller number of coefficients being stored to represent the template values. Correlation methods are widely used and applied to printed character or object shape recognition [41].

The advantage of template matching is that the pattern can be used directly without the need to devise or detect any special features contained within the pattern. The pattern image itself contains all the required information. Unfortunately this is often presented in terms of rotation, scale change, shape distortion, and noise, which makes the pattern fuzzy or indeterminate. What template matching is able to do is to define how well the pattern is matched to the template. One such definition is the *distance classifier*, given as

$$A_{m,n} = \sum_{i=0}^{N-1} \sum_{k=0}^{N-1} |x_{i,k} - t_{i-m,k-n}| \qquad (10.16)$$

where $x_{i,k}$ is the image pattern and $t_{i,k}$ is the template. It is assumed that $i - m$ and $k - n$ do not stray outside the bounds of the picture domain containing the image to be matched.

It is necessary to apply eqn (10.16) for all matches possible within the picture area and to note those coordinate values where $D_{m,n}$ is small. An extension of this definition is to apply a 2-D cross correlation between the two functions $x_{i,j}$ and $t_{i,j}$, i.e.

$$R_{pt}(m, n) = \sum_{i=0}^{N-1} \sum_{j=0}^{N-1} p_{i,j} t_{i-m,j-n} \qquad (10.17)$$

Here the template is declared similar when the cross correlation is large. This can be combined with a decision based on $R_{pt}(m, n)$ exceeding a given threshold value.

It is usual to apply eqn (10.17) through a Fourier transform invoking the equivalences with convolution (and hence correlation) noted in earlier chapters. If we let P and T represent the DFT of p and t, respectively, and consider the template function to be zero outside the picture domain, then the cross-correlation can be shown to be

$$R_{pt}(x, y) = \text{IDFT}[G_{x,y} T_{x,y}]. \qquad (10.18)$$

Although this implies that three sets of transform operations are required, it is generally easier to do this than to carry out the correlation directly.

Unless the expected signal pattern conforms to a limited range of variation it is difficult to select a good template for each pattern class. This is the case, for example with signature recognition where a syntactic approach gives better results. Template matching through comparison in the transform domain is most effective where the function series contains a number of the features of the image being matched. Thus the identification and classification of rectangular structures is one application where the non-sinusoidal functions have been found useful. Several examples are given in the literature. Chen and Seemuller have used the Walsh transform for template matching of straight-line roads, road intersections, and rectangular buildings from aerial photographs [54]. The Haar power spectrum and the non-invariant R transform have also been used for printed character recognition [55, 56].

Several of the difficulties caused by 'fuzzy' matches have been overcome by the development of adaptive 'rubber-mask' methods of template matching [57]. Here computer-simulated flexible templates are incorporated into the matching process. This applies a study of particular patterns, noting the range of variations in scale (length, width etc.) and other variations such as angle or curve applied to the main orientation of the pattern. These variations are applied in an iterative manner to modify the template to fit the natural pattern data. The method has been most effective for specific patterns such as chromosome analysis and EEG waveforms, where a body of general information is available on the range of possible shapes. It does however demand considerable extra computational requirements to carry out this iterative fitting operation in addition to the actual matching process.

10.4.2 Decision theoretic approach

In the decision theoretic approach, instead of simply matching the input pattern with the stored templates, the classification is based on selected features extraced from the input pattern. The method actually contains two operations carried out concurrently. First, the image is examined to determine the value of a set of selected *features* suitable for the kind of comparison being made. Second, the value of the features are used to *classify* the image into groups specified by a given set of features (Fig. 10.6). The template matching technique can thus be seen as a

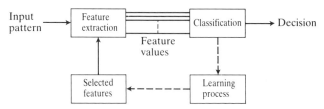

Fig. 10.6 Decision theoretic pattern recognition.

special case of decision theoretic pattern recognition, where the templates are stored in terms of feature measurement and classification consists of matching these against similar measurements taken from the input image. The decision theoretic approach can however be more sophisticated than expressed by simple template matching. An additional process shown by the dotted lines in Fig. 10.6 is to modify the classification in a dynamic fashion to improve the accuracy of recognition. This process is known as *training* or learning. It proceeds by observing patterns with known classifications so that the classifier can automatically adjust the classification weights to achieve correct recognition. The choice of such training patterns and sets of training rules to achieve convergence to an optimum set of weights for a given recognition problem are described in [58].

Classification is essentially a statistical operation, since recognition does not necessarily require a perfect representation of the input pattern. Very many forms of classification criteria are described elsewhere in terms of the application of statistical rules [59]. Two quite common methods are correlation followed by the application of a given threshold value and finding the minimum MSE between the two patterns. This latter is a variant on the distance classifier given in eqn (10.17), which uses the Euclidean distance between two vectors representing the input pattern. This is known as a *minimum-distance classifier*.

A minimum-distance classifier uses the distances between the input pattern and a set of reference values in the same domain. Let m reference vectors, R_1, R_2, . . . , R_n be associated with a given class, C_i. Then one minimum-distance classification scheme is to classify the input X as contained in class C_i,

$$X \simeq C_i \quad \text{if } |X - R_i| \text{ is a minimum.} \qquad (10.19)$$

The minimum value $|X - R_i|$ can be defined in terms of an MSE value

$$|X - R_i| = [(X - R_i)^{\mathrm{T}}(X - R_i)]^{1/2} \qquad (10.20)$$

where T is a transpose of the vector $(X - R_i)$. A full treatment of this and other statistical methods may be found in [59].

10.4.3 Syntactic approach

The decision theoretic approach discussed in the previous section is ideally suited for applications where patterns can be meaningfully represented in terms of features. Where the structure of the image plays an important role in the classification process it is less successful. It is not very effective, for example in scene analysis, since the structures and relationships of the various components in the scene are fundamental in establishing a useful pattern recognition scheme. In general the syntatic approach is used when the image patterns under consideration are so complex and the number of features needed to describe them so large that the spectrum of the image becomes too complicated to be used directly in a comparison process. Instead a hierarchical method is used in which the pattern is broken down into a series of simpler sub-patterns which can then be compared with other standard patterns in their class (Fig. 10.7). Examples are found in scene analysis, fingerprint detection and ideograph classification.

A syntactic method that is used in written character or ideograph recognition defines sub-patterns in terms of simple lines or curves, referred to as *pattern primitives,* and analysis proceeds by a set of rules known as the *pattern grammar.* Pattern primitives form a fairly simple concept, which may be used to describe a written character. For example the number 5 can be synthesized by two straight line primitives and one curved one. Unfortunately they are not too easy to define simply in terms of digitized images, and fairly complex edge following methods are used in structural analysis. Given a set of suitably defined primitives however, they may then be incorporated into a grammar or language used to describe the patterns under study. Practical techniques translate the variation in pattern into a number of 1-D sequences by dividing the picture into horizontal strips of binary characters [60]. Classification proceeds on the basis of the 0 and 1 content of a given row and a variety of statistical methods are used to count these [60, 61]. Other methods include edge-following [62] and relaxation methods [63].

A number of special languages have been proposed for the description of specific sets of patterns such as spoken language characters, chromosome images, and fingerprint patterns [59, 64]. The construction and

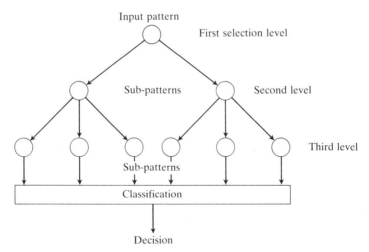

Fig. 10.7 A syntactic decision tree.

application of these languages however falls outside the scope of this book and will not be discussed further.

Transformation techniques do not play a large part in syntactic pattern recognition, which has become a province of statistical logic and compiler grammar [53]. An exception is found in the pattern recognition of Chinese ideographs or *Kanji characters,* which are composed entirely of vertical, horizontal and oblique lines. To these primitives may be added features such as corners, T-shapes, and rectangles. All of these may be classified fairly well using the Walsh or Hadamard transform [65, 66]. In Wang and Shian's implementation, the 2-D transformation of the digitized character data is followed by a classifier that puts the patterns into a range of sub-groups. The features of the groups are chosen and compared with a limited set of simple features. A minimum-distance classifier is then used for the comparison process. The recognition problem has also been considered by Takahashi and Kishi [67], who use a specialized form of the Hadamard transform for this purpose. In this transform only those coefficients or products of coefficients relating to horizontal, vertical, and oblique lines are considered. This enables a compact transform definition of a simplified form of Kanji characters to be obtained (with some loss of orthogonality).

References

1. Andrews, H. C. *Computer techniques in image processing.* Academic Press, New York (1970).
2. Huang, T. S. *Picture processing and digital filtering.* Springer, Heidelberg (1975).
3. Simon, J. C. and Rosenfeld, A. *Digital image processing and analysis.* Noordhoff, Leiden (1977).
4. Wintz, P. A. Transform coding. *IEEE Proc.* **60,** 809–19 (1972).
5. Ahmed, N., Natarajan, T., and Rao, K. R. On-image processing and a discrete cosine transform. *IEEE Trans. Comput.* **C-23,** 90–93 (1974).
6. Prosser, R. T. A multi-dimensional sampling theorem. *J. math. Anal. Applic.* **16,** 574–84 (1966).
7. Huang, T. S., Schrieber, W. F., and Tretiak, O. J. Image processing. *IEEE Proc.* **59,** 1586–609 (1971).
8. Andrews, H. C. Two-dimensional transforms. In *Picture processing and digital filtering.* (ed. T. S. Huang) pp. 21–68. Springer, Berlin (1975).
9. Connor, D. J., Brainard, R. C. and Limb, J. O. Intraframe coding for picture transmission. *IEEE Proc.* **60,** 779–800 (1972).
10. Narasimha, M. J. and Peterson, A. M. On the computation of the discrete cosine transform. *IEEE Trans. Commun.* **COM-26,** 934–6, (1978).
11. Chen, W., Smith, C. H., and Fralick, S. C. A fast computational algorithm for the discrete cosine transform. *IEEE Trans. Commun.* **COM-25,** 1004–9 (1977).
12. Hein, D. and Ahmed, N. On a real-time Walsh/Hadamard/cosine transform image processor. *IEEE Trans. electromagn. Compat.* **EMC-20,** 453–7 (1978).

13. Ghanbari, M. and Pearson, D. E. Fast cosine transform implementation for television signals. *IEEE Proc.* **129,** 59–68 (1982).
14. Srinivassan, R. and Rao, K. R. An approximation to the discrete cosine transform. *Signal Process.* **5,** 81–5 (1983).
15. Jain, A. K. Some new techniques in image processing. ONR Symposium on Current Problems in Image Science, Monterey, California (1976).
16. Yip, P., and Rao, K. R. On the computation and effectiveness of discrete sine transform. *Comput. electron. Engng* **7,** 45–55 (1980).
17. Wang, Z. de. Fast algorithm for discrete sine transform implemented by fast cosine transform. *IEEE Trans. Acoust. Speech Signal Process.* **ASSP-30,** 814–15 (1982).
18. Chen, W. H. and Smith, C. H. Adaptive coding of monochrome and color images. *IEEE Trans. Commun.* **COM-25,** 1285–92 (1977).
19. Jones, H. W. A comparison of theoretical and experimental video compression designs. *IEEE Trans. electromagn. Compat.* **EMC-21** (1), 50–6 (1979).
20. Habibi, A. and Wintz, P. A. Image coding by linear transformation and block quantization. *IEEE Trans. commun. Technol.* **COM-19,** 50–64 (1971).
21. Sakrison, D. J. and Agazi, V. R. Comparision of line by line and two-dimensional encoding of random images. *IEEE Trans. inf. —* **IT-17,** 386–98 (1971).
22. Netravoli, A. N. and Limb, J. O. Picture coding; a review. *IEEE Proc.* **68,** 366–406 (1980).
23. Jalali, A. and Rao, K. R. A high-speed FDCT processor for real-time processing of NTSC color TV signal. *IEEE Trans. electromag. Compat.* **EMC-24,** 278–86 (1982).
24. Kamagar, F. A. and Rao, K. R. Interfield hybrid coding of component color television signals. *IEEE Trans. Commun.* **COM-29,** 1740–53 (1981).
25. Frieden, B. R. Image enhancement and restoration. In *Picture processing and digital filtering* (ed. T. S. Huang) 177–248. Springer, Berlin (1975).
26. Sondhi, M. M. Image restoration: the removal of spatially invariant degradations. *IEEE Proc.* **60,** 842–53 (1972).
27. Pratt, W. K. Fast computational techniques for generalized two-dimensional Wiener filtering. *IEEE Trans. Comput.* **C-21,** 636–41 (1972).
28. Habibi, A. Two-dimensional Bayesian estimate of images. *IEEE Proc.* **60** (7), 878–83 (1972).
29. Nahi, N. E. and Assefi, T. Bayesian recursive image enhancement. *IEEE Trans. Comput* **C-21,** 734–8 (1972).
30. Helstrom, C. W. Image restoration by the method of least squares. *J. opt. Soc. Am.* **57** (3), 297–303 (1967).
31. Horner, J. L. Optical restoration of images blurred by atmospheric turbulence using optimum filter theory. *Appl. Opt.* **9,** 167–71 (1970).
32. Cannon, M. Blind deconvolution of spatially invariant image blurs with phase. *IEEE Trans. Acoust. Speech signal Process.* **ASSP-24** (1), 58–63 (1976).
33. Smith, E. G. A new two-dimensional image filtering technique based on a one-dimensional adaptive filtering method. IEEE Computer Society Conference on Pattern Recognition and Image Processing, Chicago, pp. 113–16 (1978).
34. Cappellini, V., Constantinides, A. G. and Emiliani, P. *Digital filters and their applications.* Academic Press, London (1978).

35. Hunt, B. R. Digital image processing. *IEEE Proc.* In **63,** 693–708 (1975).
36. Castan, S. Image enhancement and restoration. In *NATO ASI on Digital Image Processing and Analysis* (eds). J. C. Simon and A. Rosenfeld pp. 47–62. Noordhoff, Leiden (1977).
37. Hall, E. L., Kruger, R. P., Dwyer, S. J., Hall, D. L., and McCaren, R. W. A survey of preprocessing and feature extraction techniques for radiographic images. *IEEE Trans. Comput* **C-20,** 1032–44 (1971).
38. Wang, Z. H. A simple edge detection by two-dimensional spatial sequency digital filter. IEEE Proceedings of the International Conference on Pattern Recognition, Munich, Germany (1982).
39. Dixit, V. V. Edge extraction through Haar transform IEEE Proceedings of the Asilomar 14th Conference on Circuits, Systems, and Computers Pacific Grove, pp. 141–3 (1981).
40. Underwood, S. A. and Aggarwal, J. K. Methods of edge detection in visual scenes. IEEE Proceedings of the International Symposium on Circuit Theory, pp. 45–51, (1973).
41. Duda, R. O. and Hart, P. E. *Pattern classification and scene analysis.* Wiley, New York (1973).
42. Lynch, R. T. and Reiss, J. J. Haar transform image coding. Proceedings of the National Telecommunications Conference, Dallas, pp. 44.3.1–44.3.5 (1976).
43. Sivak, G. The Haar transform: its theory and computer implementation. Army Armament Research and Development Command, NTIS AD-AO70518/6, Washington, DC (1976).
44. Sivak, G. Applications of the Haar transform to IR imagery. Army Armament Research and Development Command, NTIS AD-AO82296/5, Washington, DC (1976).
45. Levialli, S. Finding the edge. *In Digital image processing* (ed. J. C. Simon and R. M. Haralick). Reidel, Dordrecht (1980).
46. Robinson, G. S. Edge detection by compass gradient, *Comput. Graphics image Process.* **6** (5), 492–501 (1977).
47. Shore, J. E. On the applications of Haar functions. *IEEE Trans. Commun.* **COM-21,** 209–16 (1973).
48. Andrews, H. C., Tescher, A. G., and Kruger, R. P. Image processing by digital computer. *IEEE Spectrum* **July** 20–32, (1972).
49. Ulman, J. R. *Pattern recognition techniques.* Crane, Russak (1973).
50. Tou, J. T., and Gonzalez, R. C. *Pattern recognition principles.* Addison-Wesley, New York (1974).
51. Fu, K. S., ed. Special issue *IEEE Proc. pattern Recogn. image Process.* **67,** (5), 707–856 (1979).
52. Gonzalez, R. C., and Thomason, M. C. *Syntactic pattern recognition: an introduction.* Addison-Wesley, Reading, MA, (1978).
53. Fu, K. S. *Syntactic pattern recognition and applications.* Prentice-Hall, Englewood Cliffs (1982).
54. Chen, P. F. and Seemuller, W. W. Application of Walsh transforms for topographic feature extraction using a sensory array system. *IEEE Trans. Instrum. Meas.* **IM-29,** 52–7 (1980).
55. Narasimhan, M. A. Devarajan, V., and Rao, K. R. Simulation of alphanumeric machine print recognition. *IEEE Trans. Syst. Man Cyberand* **SMC-105,** 270–75 (1980).

56. Wendling, S., Gagneux, G., and Staman, G. Use of the Haar transform and its properties in character recognition. IEEE Proceedings of the International Conference on Pattern Recognition, Boulder, Colorado, pp. 844–8 (1976).

57. Widrow, B. The 'rubber mask' technique I and II. In *Learning systems and intelligent robots* (ed. K. S. Fou and J. T. Ton). Plenum Press, New York (1974).

58. Nilsson, N. J. *Learning machines—foundations of trainable pattern classification systems*. McGraw-Hill, New York (1965).

59. Fu, K. S., ed. *Application of pattern recognition*. CRC Press, Florida (1982).

60. Ulmann, J. R. Picture analysis in character recognition. In *Digital picture analysis* (ed. A. Rosenfield) Vol. 2, pp. 56–67. Springer, Berlin (1976).

61. Ulmann, J. R. Advances in character recognition. In *Applications of pattern recognition*. (ed. K. S. Fu). pp. 162–74. CRC Press, Florida (1982).

62. Ali, F. and Pavlidis, T. Syntactic recognition of handwritten numerals. *IEEE Trans. Syst. Man Cybern.* **SMC-7,** (7), 537–41 (1977).

63. Rosenfeld, A. Relaxation methods in image processing and analysis, IEEE Fourth International Conference on Pattern Recognition, B6, Piscataway, NJ (1978).

64. Fu, K. S. Tree languages and syntactic pattern recognition. In *Pattern Recognition and Artificial Intelligence,* (ed. C. H. Chen). pp. 86–98. Academic Press, New York (1976).

65. Wang, P. P. and Shian, R. C. Machine recognition of printed Chinese characters via transformation algorithms. *Pattern Recognition* **5,** 303–21 (1973).

66. Takahashi, K. Feature extraction method by Hadamard transform. *IEEE Proc. Jpn.* Tohoku, p. 2G-8, (1980).

67. Takahashi, K. and Kishi, T. Feature extraction by specialised Hadamard transform, IEEE Proceedings of the Fifth International Conference on Pattern Recognition, pp. 1198–200 (1980).

SELECTED ADDITIONAL REFERENCES

Signal Processing

1. Agarwal, R. C. and Burrus, C. S. Fast one-dimensional digital convolution by multi-dimensional techniques. *IEEE Trans. Acoust. Speech Signal Process.* **ASSP-22,** 1–10 (1974).
2. Agarwal, R. C. and Cooley, J. C. New algorithms for digital convolution. *IEEE Trans. Acoust. Speech Signal Process.* **ASSP-25,** 392–410 (1977).
3. Allen, J. B. Short-term spectral analysis, synthesis and modification by discrete Fourier transform. *IEEE Trans. Accoust. Speech Signal Process* **ASSP-25,** 235–8 (1977).
4. Bremer, J. W. Kronecker products and matrix calculus in system theory. *IEEE Trans. Circuits Syst.* **CAS-25,** 772–81 (1979).
5. Brunn, G. Z-transform DFT filters and FFTs. *IEEE Trans. Acoust. Speech Signal Process* **ASSP-26,** 56–63 (1978).
6. Gull, S. F. and Skilling, J. Maximum entropy method in image processing. *IEEE Proc.* **131F** (6), 646–59 (1984).
7. Huange, C. H., Peterson, D. G., Rauch, H. E., Teague, J. W., and Fraser, D. F. Implementation of a fast digital processor using the residue number system. *IEEE Trans. Circuits Syst.* **CAS-28** (1), 32–38 (1981).
8. Jerri, A. J. The Shannon sampling theorem and its various extensions and applications—a tutorial review. *IEEE Proc.* **65,** 1565–96 (1977).
9. Nussbaumer, H. J. Fast polynomial transform algorithms for digital convolution. *IEEE Trans. Acoust. Speech Signal Processes* **ASSP-28,** 205–15 (1980).
10. Otnes, R. K. and Enochson, L. *Digital time series analysis.* Wiley, New York (1972).
11. Schooneveld, C. V. and Frijling, D. J. Spectral analysis: on the usefulness of linear tapering for leakage suppression. *IEEE Trans. Acoust. Speech Signal Process.* **ASSP-29,** 323–9 (1981).
12. Taylor, R. G. Stable spectral estimates with maximum entropy estimations. *IEE Electron. Lett.* **12** (50), 519–20 (1976).

Transformation

1. Bergland, G. D. A guided tour of the fast Fourier transform. *IEEE Spectrum* **6,** 41–52 (1969).
2. Blanken, J. D. and Ruston, P. L. Selection criteria for the efficient implementation of FFT algorithms. *IEEE Trans. Acoust. Speech Signal Process.* **ASSP-30,** 107–9 (1982).
3. Clarke, C. P. K. Hadamard transformation; assessment of bit-rate reduction methods. *BBC Research Report,* RD 1976/28. BBC, London (1976).
4. Cooley, J. W., Lewis, P. A. W., and Welch, P. D. Historical notes on the fast Fourier transform. *IEEE Audio Electroacoust.* **AU-15** (2), 76–9 (1967).

5. Collesidis, R. A., Dutton, T. A., and Fisher, J. R. An ultra high speed FFT processor. *IEEE Trans. Acoust. Speech Signal Process.* **ASSP-28** (1980).
6. Dickinson, B. and Steiglitz, K. An approach to the diagonalization of the discrete Fourier transform. IEEE Conference on Acoustics, Speech and Signal Processes, pp. 227–30 (1980).
7. Glassman, J. A. A generalization of the fast Fourier transform. *IEEE Trans. Comput.* **C-19** (2), 105–15 (1970).
8. Morris, L. R. A comparative study of time-efficient FFT and WFTA programs for general-purpose computers. *IEEE Trans. Acoust. Speech Signal Process.* **ASSP-26** (2), 141–50 (1978).
9. NTIS. Applications of the fast Fourier transforms. Citations from the NTIS data base, PP82–804139, Feb. (1982).
10. Patterson, R. W. and McClellen, J. H. Fixed point error analysis of Winograd Fourier transform algorithm. *IEEE Trans. Accoust. Speech Signal Process.* **ASSP-26,** 447–55 (1979).
11. Pease, M. C. Adampation of the fast Fourier transform for parallel processing. *J. Ass. Comput. Mach.* **15** (2), 252–64 (1968).
12. Pichler, F. Walsh functions and linear system theory. Proceedings of the Conference on the Applications of Walsh Functions, Washington DC, pp. 175–82 (1970).
13. Preuss, R. Very fast radix-2 FFT algorithms. *IEEE Trans. Acoust. Speech Signal Process.* **ASSP-30** (8), 595–607 (1982).
14. Rayner, P. J. W. Number theoretic transforms. In *Aspects of Signal Processing* (ed. G. Tocconi) pp. 333–53. Reidel, Dordrecht (1977).
15. Satin, P. A multi-purpose set of routines for the fast Fourier transform. *Signal Process.* **s4,** 460–2 (1982).
16. Siegman, A. E. How to compute two complex even Fourier transforms with one transform step. *IEEE Proc.* **63,** 544 (1975).
17. Slutter, J. A. Generalized running discrete transformation. *IEEE Trans. Acoust. Speech Signal Process.* **ASSP-30** (1), 60–8 (1982).

Hardware

1. Arvind, D. K., Robinson, I. W., and Parker, N. A VLSI chip for real time image processing. IEEE International Symposium on Circuits and Systems pp. 405–8 (1983).
2. Blasco, R. W. Evolution of the signal chip digital signal processing: past, present and future, IEEE International Conference on Acoustics, Speech and Signal Processing, Vol. 1, pp. 417–21 (1980).
3. Burrus, C. S. and Parkes, T. W. *DFT/FFT and convolution algorithms.* Wiley-Interscience, New York (1984).
4. Daly, D. F. and Bergeron, L. E. A programmable voice digitiser using the TI TMS320 microcomputer. IEEE International Conference on Acoustics, Speech and Signal Processing, Boston, pp. 475–8 (1983).
5. Herrman, O. E. and Smit, I. A. User-friendly environment to implement algorithms on single-chip digital signal processors. EUSIPCO '83, Erlangen, pp. 851–4 (1983).
6. Lewis, M. F., West, C. L., Deacon, J. M., and Humphries, R. F. Recent developments in SAW devices. *IEEE Proc.* **131A,** 186–215 (1984).

7. McIlroy, C. M., Linggard, R., and Monteith, W. Hardware for real-time image processing. *IEE Proc.* **131E** (6), 223–9 (1984).
8. Swartzlander, E. E. Signal processing architecture with VLSI. IEEE International Conference on Acoustics, Speech and Signal Processing, pp. 369–70 (1980).
9. Thompson, J. S. and Tewksby, S. K. LSI signal processing architecture for telecom applications. *IEEE Trans. Acoust. Speech Signal Process.* **ASSP-30,** 613–31 (1984).
10. Waser, S. Survey of VLSI for digital signal processing. IEEE International Conference on Acoustics, Speech and Signal Processing, pp. 377–80 (1980).
11. Winter, G. E. and Yamashita, R. R. A single-board floating-point signal processor. IEEE International Conference on Acoustics, Speech and Signal Processing, Boston, pp. 947–50 (1983).

Applications

1. Andrews, H. *Introduction to mathematical techniques in pattern recognition.* Prentice-Hall, Englewood Cliffs (1972).
2. Burdic, W. S. *Radar signal analysis.* Prentice-Hall, Englewood Cliffs (1968).
3. Feldman, J. A. Hofsletter, E. M., and Malpass, M. L. A compact flexible LPC vocoder based on a commercial signal processor microcomputer. *IEEE Trans. Acoust. Speech and Signal Process.* **ASSP-31** (1), 252–7 (1983).
4. Gorguir, N. Comparative performance of SVD and adaptive cosine transform in coding images. *IEEE Trans. Commun.* **COM-27,** 1230–4 (1979).
5. Hagiwari, P., Kita, Y., Miyamoto, T., Toba, Y., Hara, H. and Akazara, T. A high-performance signal processor for speech synthesis and analysis. IEEE International Conference on Acoustics, Speech and Signal Processing, Paris SP8–11 (1982).
6. Kamangar, F. A fast algorithm for the 2-D discrete cosine transform. *IEEE Trans. Comput.* **C-31** (9), 899–906 (1977).
7. Knight, W. C., Pridham, R. G., and Kay, S. M. Digital signal processing for sonar. *IEEE Proc.* **69,** 1451–506 (1981).
8. Levy, J. B. and Linkens, D. A. Spectral analysis of short-time biomedical data using adaptive filters. *IEEE Proc.* **131A** (3), 164–9 (1984).
9. Morgera, S. D. Digital signal processor for precision wide swath bathymetry. *IEEE J. oceanic Engng* **OE-9** (2), 73–84 (1984).
10. Ngam, K. N. Adaptive transform coding of video signals. *IEE Proc.* **129E** (1), 28–40 (1982).
11. Oppenheim, A. V., ed. *Applications of digital signal processing.* Prentice-Hall, Englewood Cliffs (1978).
12. Richards, M. A. Helium speech enhancement using the short-term Fourier transform. *IEEE Trans. Acoust. Speech Signal Process.* **ASSP-30** (6), 841–53 (1982).
13. Trider, R. C. A fast Fourier transform-based sonar signal processor IEEE International Conference on Acoustics, Speech and Signal Processing, pp. 389–93 (1976).

INDEX

absorbtivity factor 227
acoustic analysis 179
active sonar 194, 197
adaptive
  coding compression 242
  modulator 197
  threshold 207
  transform coding 167
alaised signals 5, 239; *see also* principal
  alais
algorithm
  convolution 99
  Cooley–Tukey 69, 70, 71, 76
  fast Fourier transform 77
  fast Haar transform 114
  fast Walsh transform 10
  in-place 85
  Larsen's 113
  Manz's 113
  mixed radix 70, 91
  number theoretic 94
  prime factor 94
  Rader's 99
  radix 4 70, 91
  radix 8 70, 91
  reduced multiplication 93
  Sande–Tukey 7
  short DFT 98
  twiddle factor 88
  Winograd 94, 143
analog signal 2
anti-aliasing filter 7
aperture 5
  error 5
array
  linear 200
  NORSAR seismic 207
  processing 133
  programmable logic 146
  seismic 204
  sonar 289
  transducer 197
  two-dimensional 45
assembler code 158
assembly language 154
atomic spin 230
auto-correlation 17, 172, 209
  coefficient 23

auto-regressive model 22
  moving average model 22, 223

back-projection integral equation 228
band-limited function 6
bandwidth compression 219
Barker code 209
Bartlett window 13
beam-forming 197, 205
binary-weighted series 75
biomedical processing 218
biomedicine 218
bit-reversed notation 65
bit slice operation 133
  architecture 133
Blackman window 13
Blackman–Tukey method 20, 21
block circulant matrix 64
block-diagonal matrix 121
burst waveform 209
butterfly 85
  computation 78
  cosine/sine 128
  loops 110
  module 136, 143

CAL function 36
carrier-free radar 213
Cartesian coordinate 4, 202, 228
CCD, *see* charge-coupled device
CCD Walsh transform 148
       cepstral homomorphic processing 173
cepstrum 24, 172, 196
  power 196
channel vocoder 168
characteristic roots 57
charge-coupled device 146
Chinese ideographs 256
Chinese remainder theorem 96
chirp filter 147
  signal 18, 209
  -Z SAW filter 149
  -Z transform 70, 147, 149
  -Z sliding transform 148
circulant matrix 61, 63

circular convolution 64, 99
circular correlation 15
C-matrix transform 242
coding
  adaptive transform 167
  assembler 158
  Barker 209
  intraframe 241, 244
  linear predictive 170
  mask 46
  predictive 184
  pseudo-random 198
  run-length 183
  source 165
  speech 165
  sub-band 166
  transform 184
column vector 49
communications 163
complex
  cepstrum 204
  conjugate 30, 50
  correlation 199
  Fourier series 6, 29, 30
computer-assisted tomography 226
conjugate symmetry 43, 89
constant geometry 110, 138
continuous
  signal 2
  speech 179
contour edge map 250
contour tracking 248
convolution
  algorithm 99
  back-projection 227, 232
  see also correlation
Cooley–Tukey algorithm 69, 71, 76
correlation 14
  auto 17, 172, 209
  circular 15
  coefficient 14
  complex 199
  cross 14, 16, 195
  delay 14
correlogram 16
cosine matrix 128
cosine/sine butterfly 128
cosine transform 44
cross-correlation 14, 16, 195
  coefficient 14
cross-correlogram 16
C–T, see Cooley–Tukey
C–T algorithm 69, 71, 76

data
  acquisition 5
  communication 180

compression 183, 238, 244
compression ratio 183, 221
modelling 22
decimation in
  frequency 70, 79
  time 70, 78
decision theoretic method 251, 253
deconvolution 203
delay line 147
  time compression 199
delta
  function 231
  modulation 166
DELTIC 199
determinants 55
deterministic signal 2
DFT, see discrete Fourier transform
diagonal matrix 53
differential pulse code modulation 244
digital
  filtering 8, 147
  processing 2
  sequency division multiplexing 185
  signal processor 154
digitization 238, 240
digit recognition 178
Dirac function 2, 32
direct memory access 144
discrete
  cosine transform 44, 148, 167, 242
  Fourier transform 19, 33, 34, 77
  Fourier transform properties 34
  Haar transform 39
  Walsh matrix 108
  Walsh series 36
  Walsh transform 39, 109
dispersive delay line 149
distance classifier 252
dominant rythms 222
Doppler
  equation 209
  processing 212
  shift 199
  spreading 199
double-sideband signal 181
dyadic
  matrix 61, 62
  order 109

earthquake analysis 202
ECG, see electrocardiography
ECG analysis 219
edge detection 248
EEG, see electroencephalography
EEG analysis 222
eigenvalues 56
eigenvectors 57

electrocardiography 218
electroencephalography 218
emission-computed tomography 225, 232
explosion seismology 202
exponential function 32

factorising, *see* sparse matrix factorizing
fan-beam geometry 229
fast
  cosine transform 127
  Fourier transform 77
  Haar transform 114
  Hadamard transform 109
  slant transform 118
  transformation 109
  unitary transform 124
  Walsh transform 10
FFT, *see* fast Fourier transform
FFT spectrum 222
FHT, *see* fast Hadamard transform
filter
  alaising 7
  bank 188
  chirp 147
  chirp Z 149
  digital 8, 147
  finite impulse response 9
  FIR 9
  SAW 149
  IIR 9, 171
  inverse 246
  matrix 13
  multiple digital 156
  non-recursive 9
  recursive 9
  threshold 246
  time 173
  transversal 147
  weights 9
  Wiener 13, 40, 211, 225, 246
FIR, *see* finite impulse response filter
FIR filter 9
fixed radix transform 68
flow diagram, *see* signal flow diagram
formant
  classification 165
  synthesis 176
  vocoder 169
Fourier
  coefficients 27
  complex series 6, 29, 30
  discrete transform 33, 34, 77
  fast transform 77
  inverse discrete transform 32, 34
  odd-time, odd-frequency transform 187
  parallel transform 138
  series 3, 27

  spectrum 28, 222
  transform 3, 10, 11, 19, 31, 227
frequency
  division multiplexing 180
  transfer function 246
functions 48
fuzzy matching 253

Gaussian transfer function 247
generalized
  matrix 120
  transform 120
  Wiener filter 13
Gibbs phenomenon 11
gradient operator 248
gray code 66
  conversion 108
grey levels 251

Haar
  discrete transform 39
  function series 35, 38
  transform 39, 249
Hadamard
  matrix 61, 108, 109
  ordering 108
  SAW transform 150
  series 109
Hamming window 12
Hanning window 12, 21
hardware 132
  transform 132
Harvard architecture 155
Hermitian operator 52
homomorphic
  deconvolution 203
  processing 172
  vocoder 172

identity operator 52
IIR, *see* infinite impulse response filter
IIR filter 9
ill-conditioned process 245
image
  compression 130, 241, 249
  enhancement 237, 245
  matrix 237, 238
  processing 4, 218, 237
  reconstruction 243
  restoration 237, 245
  transmission 241, 242
impulse
  radar 213
  sequence 2
index shuffling 92

infinite impulse response filter 9, 171
inner product 49
in-place
    algorithm 85
    Fourier transform 85
    Haar transform 117
    Walsh transform 111
intensity mapping 248
interframe coding 244
into-the-ground radar 213
intraframe coding 241, 244
inverse
    filter 246
    Fourier transform 32, 34
    transformation 88
isotope decay 233

Kanji characters 256
Karhunan–Loève
    series 35, 57
    transform 44, 57, 68, 120, 221
kernel 52
Kronecker
    multiplication 59
    ordering 108
    product 57

lag
    function 14
    values 23
Laguerre polynomials 35
Larmor frequency 230
Larsen's algorithm 113
least squares fitting 23
left-diagonal matrix 54
Legendre function 35
limited bandwidth image 243
linear
    array 200
    difference 9
    linear equation 171
    isoplanar system 241
    mathematics 48
    operator 51
    prediction 170
    predictive coding 170
    predictive method 22
linguistic
    grammar 224
    model 169, 251
'look ahead' operation 154

majority logic multiplexing 185
Manz's algorithm 113
masking 250
mask coding techniques 46

matched filter 211
matrix 51
    algebra 48, 52
    array 45
    block circulant 64
    block diagonal 121
    circulant 61, 63
    C 242
    diagonal 53
    dyadic 61, 62
    factorization 58
    generalized 120
    Hadamard 61, 108, 109
    image 237, 243
    left-diagonal 54
    orthogonal 55, 99
    permutation 80
    rectangular 103
    right-diagonal 54, 81
    shuffling 80
    slant 119
    sparse 107
    specialized 61
    symmetrical 54
    transpose 54
    unit 52, 54
    unitary 55
    Walsh 107
maximum
    entropy 22, 212
    likelihood method 22
mean-square-error 14
microprocessor implementation 141
microprogramming 134, 154
microtechnology 144
minimum distance
    classifier 254
    score 178
mixed radix algorithm 70, 91
modem 182
modular architecture 152
modulation transfer function 240
modulo-2
    addition 66
    arithmetic 65
modulus 50
Mohovoric discontinuity 203
Moire patterns 239, 247
multi-dimensional processing 218
multi-dimensional signal 4
multiple digital filter 156

natural order 109
non-recursive filter 9
non-singular operator 53
non-sinusoidal
    function 35
    transform 14, 107

non-stationary
   radar 212
   signal 3, 222
normalized Hadamard matrix 62
NORSAR seismic array 207
nuclear magnetic resonance 225, 229
number-theoretic
   algorithm 94
   transform 68, 70, 94
Nyquist frequency 6

odd-time, odd-frequency FFT 187
orthogonal 27, 37, 51
   matrix 55, 99
   transform 243
orthonormal 27, 51
oscillatory response 12
overlap-add 10

parameter extraction 223
parametric method 20
parallel FFT 138
   processing 137, 135
parse trees 224
Parsival's theorem 29
passive sonar 194
pattern
   grammar 255
   primitives 255
   recognition 251, 255
perfect shuffle 123, 138
periodogram 20, 41, 203
permutation matrix 80
phase-invariant transform 114
phase transfer function 240
phonemes 165
phonetic analysis 180
picture element 24, 227, 237
pipeline processing 133, 139, 156
pitch
   frequency 165
   period 170
   recognition vocoder 170
pixel 24, 41, 227
point spread function 240
positive emission tomography 233
post-weave operation 102
power
   cepstrum 196
   spectral density 19
   spectrum 23, 173
precession 230
predictive
   coding 184
   deconvolution 203
pre-weave operation 101
prime factor algorithm 94

prime number factorizing 94
principal alais 5
programmable
   logic array 146
   signal processor 152
projection theorem 228
prosodic analysis 167
pseudo-colour 251
pseudo-random code 198
PSP, *see* programmable signal processor
pulse-coded modulation 181

QRS cycle 219
quantization 2, 7, 239
quantizer 174

radar
   applications 193, 208
   signal processing 18
Rademacher
   function 109
   series 213
Rader's algorithm 99
radio
   frequency spectrum 230
   isotope scanning 225
radix
   4 algorithm 70, 91
   8 algorithm 70, 91
   2 transform 74
   r transform 70
random signals 3
range spreading 199
rapid transform 114
rationalized Haar transform 115
real-valued function 12
rectangular
   matrix 103
   pulse 32
recursive filter 9
reduced multiplication algorithm 93
relative bandwidth 212
residue modulo-b 50
right-diagonal matrix 54, 81
rotational transformation 227
round-off noise 68
row column method 127
row vector 49
'rubber mask' technique 253
run-length coding 183

SAL function 36
sampling 5, 243
   function 243
   interval 2, 5
   theorem 6

Sande–Tukey algorithm 77
SAW, *see* surface acoustic wave
scalar values 48
scaling factor 34
scanning 225
scintiography 225
segmentation method 20, 22
seismic
    array 204
    data compression 207
    signal analysis 203
seismogram 202
seismology 193, 202
select save 10
sequency 36
    bandlimiting 207
    multiplexing 185
shift theorem 33, 40
short DFT algorithms, *see* small-N
    algorithms
shuffling matrix 80
sidebands 181
signal flow diagram 79, 83, 103, 110
signal processor 152
signal-to-noise ratio 8, 14, 17, 140, 168
sinc function 32
single-chip signal processor 152, 154; *see
    also* programmable signal processor
single-dimensional signal 2, 218
single-sideband signal 180
singular
    operator 52
    value decomposition 241
slant
    matrix 119
    transform 56
sliding chirp-Z transform 148
small-N algorithm 98, 103
Sobel operator 249
software implementation 143
sonar
    applications 193
    array 197
    imaging 200
    processing 193
source coding 165
sparse matrix factorizing 107
specialized matrices 61
spectral
    analyser 156
    analysis 19, 147
    window 10
speech
    coding 165
    processing 163, 164
    recognition 176
    synthesis 175
    synthesizer 175

S–T, *see* Sande–Tukey
S–T algorithm 77
stored spectra 177
sub-band coding 166
sub-optimal transform 68
successive doubling method 77
surface acoustic wave 148
swept frequency signal 18, 198
symmetrical matrix 54
syntactive method 223, 251, 254

target location 208
TDM–FDM conversion 186
television
    data compression 244
    image 241
template 177
template matching 251, 253
threshold
    detection 18
    filter 246
time filtering 173
time-division multiplexing 181
tomography 218, 225
    computer assisted 225, 226
    emission computeed 225, 232
    positive emission 233
training in pattern recognition 254
transducer array 197
transfer function 246, 247
transformation of real data 89
transform
    adaptive 167
    CCD Walsh 148
    chirp Z 70, 147, 149
    coding 184
    cosine 44
    C matrix 242
    discrete cosine 44, 148, 167, 242
    discrete Fourier 33, 34, 77
    discrete Haar 39
    discrete Walsh 39, 109
    fast 109
    fast cosine 127
    fast Fourier 77
    fast Haar 118
    fast Hadamard 109
    fast slant 118
    fast unitary 124
    fixed radix 68
    Fourier 3, 10, 11, 19, 31, 227
    generalized 120
    Haar 39, 249
    Hadamard–SAW 150
    inverse 32
    inverse Fourier 32, 34
    Karhunan–Loève 44, 57, 68, 120, 221

odd-time, odd-frequency 187
orthogonal 243
parallel 138
radix 2 70
radix r 70
rapid 114
rationalized Haar 115
slant 56
sliding chirp Z 149
sub-optimal 68
two-dimensional 44, 45, 125, 227
unitary 122
VLSI 144
Winograd 101
Walsh 39, 207, 242
Z 173
transmultiplexer 182, 186
transpose of a matrix 54
transversal filter 147
triangular window 13
truncation 11
twiddle factor 78
  algorithm 88
two-dimensional
  array 45
  Fourier transform 44, 227
  Haar function 46
  processing 24, 224, 237
  SAW transform 150
  transform 41, 125
  Walsh transform 45
two-level factorizing 71
two-pulse canceller 209
two-sided spectra 7

ultrasonic imaging 225, 231
ultrasound 231
unit matrix 52, 54
unitary
  matrix 55
  transform 122

vectors 48
videotext 242
VLSI
  chip technology 163
  transform 144
vocoder 165, 168
voiced/unvoiced speech 179

Walsh
  carrier 213
  discrete transform 107
  function series 35
  Hadamard fast transform 111
  matrix 52, 108
  series 213
  transform 39, 207, 242
  waves 213
waveforms 1
Weaver modulation 186
white noise 8, 22, 172, 206
wide-band noise 17
Wiener filtering 13, 40, 211, 225, 246
Wiener–Khintchine theorem 21, 35
window
  Blackman 13
  Bartlett 13
  function 11, 20
  Hamming 13
  Hanning 12, 21
  spectral 10
  triangular 13
Winograd
  algorithm 94, 143
  transform 101
word recognizer 177
word spotting 180

X-ray detector 226

Z-transform 173
zero-crossing rate 178; *see also* sequence